T0291806

CAMBRIDGE LIBRARY COLLECTION

Books of enduring scholarly value

Physical Sciences

From ancient times, humans have tried to understand the workings of the world around them. The roots of modern physical science go back to the very earliest mechanical devices such as levers and rollers, the mixing of paints and dyes, and the importance of the heavenly bodies in early religious observance and navigation. The physical sciences as we know them today began to emerge as independent academic subjects during the early modern period, in the work of Newton and other 'natural philosophers', and numerous sub-disciplines developed during the centuries that followed. This part of the Cambridge Library Collection is devoted to landmark publications in this area which will be of interest to historians of science concerned with individual scientists, particular discoveries, and advances in scientific method, or with the establishment and development of scientific institutions around the world.

Conversations on Chemistry

Jane Haldimand Marcet (1769–1858) was a pioneer in the field of education who wrote accessible introductory books on science and economics. Noting that women's education 'is seldom calculated to prepare their minds for abstract ideas', she resolved to write books that would inform, entertain and improve a generation of female readers. First published anonymously in 1805, her two-volume work *Conversations on Chemistry* swiftly became a standard primer, going through sixteen editions in England alone, and was cited by Michael Faraday as having greatly influenced him. Presented as a series of discussions between a fictional tutor, Mrs Bryan, and her two female students, the flighty Caroline and earnest Emily, *Conversations* combines entertaining banter with a clear and concise explanation of scientific theories. In Volume 1 the girls are introduced to 'Simple Bodies' through such colourful examples as hot air balloons and the spa waters of Harrogate. For more information on this author, see http://orlando.cambridge.org/public/svPeople?person_id=marcja

Cambridge University Press has long been a pioneer in the reissuing of out-of-print titles from its own backlist, producing digital reprints of books that are still sought after by scholars and students but could not be reprinted economically using traditional technology. The Cambridge Library Collection extends this activity to a wider range of books which are still of importance to researchers and professionals, either for the source material they contain, or as landmarks in the history of their academic discipline.

Drawing from the world-renowned collections in the Cambridge University Library, and guided by the advice of experts in each subject area, Cambridge University Press is using state-of-the-art scanning machines in its own Printing House to capture the content of each book selected for inclusion. The files are processed to give a consistently clear, crisp image, and the books finished to the high quality standard for which the Press is recognised around the world. The latest print-on-demand technology ensures that the books will remain available indefinitely, and that orders for single or multiple copies can quickly be supplied.

The Cambridge Library Collection will bring back to life books of enduring scholarly value (including out-of-copyright works originally issued by other publishers) across a wide range of disciplines in the humanities and social sciences and in science and technology.

Conversations on Chemistry

In which the Elements of that Science are Familiarly Explained and Illustrated by Experiments

VOLUME 1:
ON SIMPLE BODIES

JANE HALDIMAND MARCET

CAMBRIDGE UNIVERSITY PRESS

Cambridge, New York, Melbourne, Madrid, Cape Town, Singapore,
São Paolo, Delhi, Dubai, Tokyo, Mexico City

Published in the United States of America by Cambridge University Press, New York

www.cambridge.org
Information on this title: www.cambridge.org/9781108016834

This edition first published 1817
This digitally printed version 2010

ISBN 978-1-108-01683-4 Paperback

CONVERSATIONS

ON

CHEMISTRY;

IN WHICH

THE ELEMENTS OF THAT SCIENCE

ARE

FAMILIARLY EXPLAINED

AND

ILLUSTRATED BY EXPERIMENTS.

IN TWO VOLUMES.

The Fifth Edition, revised, corrrected, and considerably enlarged.

VOL. I.

ON SIMPLE BODIES.

LONDON:
PRINTED FOR LONGMAN, HURST, REES, ORME, AND BROWN,
PATERNOSTER-ROW.
1817.

Printed by A. Strahan,
Printers-Street, London.

ADVERTISEMENT.

The Author, in this fifth edition, has endeavoured to give an account of the principal discoveries which have been made within the last four years in Chemical Science, and of the various important applications, such as the gas-lights, and the miner's-lamp, to which they have given rise. But in regard to doctrines or principles, the work has undergone no material alteration.

London, July, 1817.

PREFACE.

In venturing to offer to the public, and more particularly to the female sex, an Introduction to Chemistry, the author, herself a woman, conceives that some explanation may be required; and she feels it the more necessary to apologise for the present undertaking, as her knowledge of the subject is but recent, and as she can have no real claims to the title of chemist.

On attending for the first time experimental lectures, the author found it almost impossible to derive any clear or satisfactory information from the rapid demonstrations which are usually, and perhaps necessarily, crowded into popular courses of this kind. But frequent opportunities having afterwards occurred of conversing with a friend on the subject of chemistry, and of repeating a variety of experiments, she became better acquainted with the principles of that science, and began to feel highly interested in its pursuit. It was then that she perceived, in attending the excellent lectures delivered at the Royal Institution, by the present Professor of Chemistry, the great advantage which her previous knowledge of the subject, slight as it was, gave her over others who had not enjoyed the same means of private instruction. Every fact or experiment attracted her attention, and served to explain some theory to which she was not a total stranger; and she had the gratification to find that the numerous and elegant illustrations, for which that school is so much distinguished, seldom failed to produce on her mind the effect for which they were intended.

Hence it was natural to infer, that familiar conversation was, in studies of this kind, a most useful auxiliary source of information; and more especially to the female sex, whose education is seldom calculated to prepare their minds for abstract ideas, or scientific language.

As, however, there are but few women who have access to this mode of instruction; and as the author was not acquainted with any book that could prove a substitute for it, she thought that it might be useful for beginners, as well as satisfactory to herself, to trace the steps by which she had acquired her little stock of chemical knowledge, and to record, in the form of dialogue, those ideas which she had first derived from conversation.

But to do this with sufficient method, and to fix upon a mode of arrangement, was an object of some difficulty. After much hesitation, and a degree of embarrassment, which, probably, the most competent chemical writers have often felt in common with the most superficial, a mode of division was adopted, which, though the most natural, does not always admit of being strictly pursued — it is that of treating first of the simplest bodies, and then gradually rising to the most intricate compounds.

It is not the author's intention to enter into a minute vindication of this plan. But whatever may be its advantages or inconveniences, the method adopted in this work is such, that a young pupil, who should occasionally recur to it, with a view to procure information on particular subjects, might often find it obscure or unintelligible; for its various parts are so connected with each other as to form an uninterrupted chain of facts and reasonings, which will appear sufficiently clear and consistent to those only who may have patience to go through the whole work, or have previously devoted some attention to the subject.

It will, no doubt, be observed, that in the course of these Conversations, remarks are often introduced, which appear much too acute for the young pupils, by whom they are supposed to be made. Of this fault the author is fully aware. But, in order to avoid it, it would have been necessary either to omit a variety of useful illustrations, or to submit to such minute explanations and frequent repetitions, as would have rendered the work tedious, and therefore less suited to its intended purpose.

In writing these pages, the author was more than once checked in her progress by the apprehension that such an attempt might be considered by some, either as unsuited to the ordinary pursuits of her sex, or ill-justified by her own recent and imperfect knowledge of the subject. But, on the one hand, she felt encouraged by the establishment of those public institutions, open to both sexes, for the dissemination of philosophical knowledge, which clearly prove that the general opinion no longer excludes women from an acquaintance with the elements of science; and, on the other, she flattered herself that whilst the impressions made upon her mind, by the wonders of Nature, studied in this new point of view, were still fresh and strong, she might perhaps succeed the better in communicating to others the sentiments she herself experienced.

The reader will soon perceive, in perusing this work, that he is often supposed to have previously acquired some slight knowledge of natural philosophy, a circumstance, indeed, which appears very desirable. The author's original intention was to commence this work by a small tract, explaining, on a plan analogous to this, the most essential rudiments of that science. This idea she has since abandoned; but the manuscript was ready, and might, perhaps, have been printed at some future period, had not an elementary work of a similar description, under the title of "Scientific Dialogues," been pointed out to her, which, on a rapid perusal, she thought very ingenious, and well calculated to answer its intended object.

CONTENTS

OF

THE FIRST VOLUME.

ON SIMPLE BODIES.

ERRATA.

Vol. I. page 56. last line but one, for "ealoric," read "calorific."
179. Note, for "Plate XII." r. "Plate XIII."

CONVERSATIONS
ON
CHEMISTRY.

CONVERSATION I.

ON THE GENERAL PRINCIPLES OF CHEMISTRY.

MRS. B.

AS you have now acquired some elementary notions of NATURAL PHILOSOPHY, I am going to propose to you another branch of science, to which I am particularly anxious that you should devote a share of your attention. This is CHEMISTRY, which is so closely connected with Natural Philosophy, that the study of the one must be incomplete without some knowledge of the other; for, it is obvious that we can derive but a very imperfect idea of bodies from the study of the general laws by which they are governed, if we remain totally ignorant of their intimate nature.

CAROLINE.

To confess the truth, Mrs. B., I am not disposed to form a very favourable idea of chemistry, nor do I expect to derive much entertainment from it. I prefer the sciences which exhibit nature on a grand scale, to those that are confined to the minutiæ of petty details. Can the studies which we have lately pursued, the general properties of matter, or the revolutions of the heavenly bodies, be compared to the mixing up of a few insignificant drugs? I grant, however, there may be entertaining experiments in chemistry, and should not dislike to try some of them: the distilling, for instance, of lavender, or rose water......

MRS. B.

I rather imagine, my dear Caroline, that your want of taste for chemistry proceeds from the very limited idea you entertain of its object. You confine the chemist's laboratory to the narrow precincts of the apothecary's and perfumer's shops, whilst it is subservient to an immense variety of other useful purposes. Besides, my dear, chemistry is by no means confined to works of art. Nature also has her laboratory, which is the universe, and there she is incessantly employed in chemical operations. You are surprised, Caroline; but I assure you that the most wonderful and the most interesting phenomena of nature are almost all of them produced by chemical powers. What Bergman, in the introduction to his history of chemistry, has said of this scienee, will give you a more just and enlarged idea of it. The knowledge of nature may be divided, he observes, into three periods. The first was that in which the attention of men was occupied in learning the external forms and characters of objects, and this is called *Natural History*. In the second, they considered the effects of bodies acting on each other by their mechanical power, as their weight and motion, and this constitutes the science of *Natural Philosophy*. The third period is that in which the properties and mutual action of the elementary parts of bodies was investigated. This last is the science of CHEMISTRY, and I have no doubt you will soon agree with me in thinking it the most interesting.

You may easily conceive, therefore, that without entering into the minute details of practical chemistry, a woman may obtain snch a knowledge of the science as will not only throw an interest on the common occurrences of life, but will enlarge the sphere of her ideas, and render the contemplation of nature a source of delightful instruction.

CAROLINE.

If this is the case, I have certainly been much

mistaken in the notion I had formed of chemistry. I own that I thought it was chiefly confined to the knowledge and preparation of medicines.

MRS. B.

That is only a branch of chemistry which is called Pharmacy; and, though the study of it is certainly of great importance to the world at large, it belongs exclusively to professional men, and is therefore the last that I should advise you to pursue.

EMILY.

But, did not the chemists formerly employ themselves in search of the philosopher's stone, or the secret of making gold?

MRS. B.

These were a particular set of misguided philosophers, who dignified themselves with the name of Alchemists, to distinguish their pursuits from those of the common chemists, whose studies were confined to the knowledge of medicines.

But, since that period, chemistry has undergone so complete a revolution, that, from an obscure and mysterious art, it is now become a regular and beautiful science, to which art is entirely subservient. It is true, however, that we are indebted to the alchemists for many very useful discoveries, which sprung from their fruitless attempts to make gold, and which, undoubtedly, have proved of infinitely greater advantage to mankind than all their chimerical pursuits.

The modern chemists, instead of directing their ambition to the vain attempt of producing any of the original substances in nature, rather aim at analysing and imitating her operations, and have sometimes succeeded in forming combinations, or effecting decompositions, no instances of which occur in the chemistry of Nature. They have little reason to regret their inability to make gold, whilst, by their innumerable inventions and discoveries, they have so greatly stimulated industry and facilitated labour, as prodigiously to increase the luxuries as well as the necessaries of life.

EMILY.

But, I do not understand by what means chemistry can facilitate labour; is not that rather the province of the mechanic?

MRS. B.

There are many ways by which labour may be rendered more easy, independently of mechanics; but even the machine, the most wonderful in its effects, the Steam-engine, cannot be understood without the assistance of chemistry. In agriculture, a chemical knowledge of the nature of soils, and of vegetation, is highly useful; and, in those arts which relate to the comforts and conveniences of life, it would be endless to enumerate the advantages which result from the study of this science.

CAROLINE.

But, pray, tell us more precisely in what manner the discoveries of chemists have proved so beneficial to society?

MRS. B.

That would be an injudicious anticipation; for you would not comprehend the nature of such discoveries and useful applications, as well as you will do hereafter. Without a due regard to method, we cannot expect to make any progress in chemistry. I wish to direct your observations chiefly to the chemical operations of Nature; but those of Art are certainly of too high importance to pass unnoticed. We shall therefore allow them also some share of our attention.

EMILY.

Well, then, let us now set to work regularly. I am very anxious to begin.

MRS. B.

The object of chemistry is to obtain a knowledge of the intimate nature of bodies, and of their mutual action on each other. You find therefore,

Caroline, that this is no narrow or confined science, which comprehends every thing material within our sphere.

CAROLINE.

On the contrary, it must be inexhaustible; and I am a loss to conceive how any proficiency can be made in a science whose objects are so numerous.

MRS. B.

If every individual substance were formed of different materials, the study of chemistry would, indeed, be endless; but you must observe that the various bodies in nature are composed of certain elementary principles, which are not very numerous.

CAROLINE.

Yes; I know that all bodies are composed of fire, air, earth, and water; I learnt that many years ago.

MRS. B.

But you must now endeavour to forget it. I have already informed you what a great change chemistry has undergone since it has become a regular science. Within these thirty years especially, it has experienced an entire revolution, and it is now proved, that neither fire, air, earth, nor water, can be called elementary bodies. For an elementary body is one that has never been decomposed, that is to say, separated into other substances; and fire, air, earth, and water, are all of them susceptible of decomposition.

EMILY.

I thought that decomposing a body was dividing it into its minutest parts. And if so, I do not understand why an elementary substance is not capable of being decomposed, as well as any other.

MRS. B.

You have misconceived the idea of *decomposition*; it is very different from mere *division*. The latter simply reduces a body into parts, but the former separates it into the various ingredients, or materials, of which it is composed. If we were to take a loaf of bread, and separate the several ingredients of which it is made, the flour, the yeast, the salt, and the water, it would be very different from cutting or crumbling the loaf into pieces.

EMILY.

I understand you now very well. To decompose a body is to separate from each other the various elementary substances of which it consists.

CAROLINE.

But flour, water, and other materials of bread,

according to our definition, are not elementary substances?

MRS. B.

No, my dear; I mentioned bread rather as a familiar comparison, to illustrate the idea, than as an example.

The elementary substances of which a body is composed are called the *constituent* parts of that body; in decomposing it, therefore, we separate its constituent parts. If, on the contrary, we divide a body by chopping it to pieces, or even by grinding or pounding it to the finest powder, each of these small particles will still consist of a portion of the several constituent parts of the whole body: these are called the *integrant* parts; do you understand the difference?

EMILY.

Yes, I think, perfectly. We *decompose* a body into its *constituent* parts; and *divide* it into its *integrant* parts.

MRS. B.

Exactly so. If therefore a body consists of only one kind of substance, though it may be divided into its integrant parts, it is not possible to decompose it. Such bodies are therefore called *simple* or *elementary*, as they are the elements of which all other bodies are composed. *Compound bodies* are such as consist of more than one of these elementary principles.

CAROLINE.

But do not fire, air, earth, and water, consist, each of them, but of one kind of substance?

MRS. B.

No, my dear; they are every one of them susceptible of being separated into various simple bodies. Instead of four, chemists now reckon upwards of forty elementary substances. The existence of most of these is established by the clearest experiments; but, in regard to a few of them, particularly the most subtle agents of nature, *heat*, *light*, and *electricity*, there is yet much uncertainty, and I can only give you the opinion which seems most probably deduced from the latest discoveries. After I have given you a list of the elementary bodies, classed according to their properties, we shall proceed to examine each of them separately, and then consider them in their combinations with each other.

Excepting the more general agents of nature, heat, light, and electricity, it would seem that the simple form of bodies is that of a metal.

CAROLINE.

You astonish me! I thought the metals were only

one class of minerals, and that there were besides, earths, stones, rocks, acids, alkalies, vapours, fluids, and the whole of the animal and vegetable kingdoms.

MRS. B.

You have made a tolerably good enumeration, though I fear not arranged in the most scientific order. All these bodies, however, it is now strongly believed, may be ultimately resolved into metallic substances. Your surprise at this circumstance is not singular, as the decomposition of some of them, which has been but lately accomplished, has excited the wonder of the whole philosophical world.

But to return to the list of simple bodies — these being usually found in combination with oxygen, I shall class them according to their properties when so combined. This will, I think, facilitate their future investigation.

EMILY.

Pray what is oxygen?

MRS. B.

A simple body; at least one that is supposed to be so, as it has never been decomposed. It is always found united with the negative electricity. It will be one of the first of the elementary bodies whose properties I shall explain to you, and, as you will soon perceive, it is one of the most important in nature; but it would be irrelevant to enter upon this subject at present. We must now confine our attention to the enumeration and classification of the simple bodies in general. They may be arranged as follows:

CLASS I.

Comprehending the imponderable agents, viz.

HEAT or CALORIC,
LIGHT,
ELECTRICITY.

CLASS II.

Comprehending agents capable of uniting with inflammable bodies, and in most instances of effecting their combustion.

OXYGEN,
CHLORINE,
IODINE.

CLASS III.

Comprehending bodies capable of uniting with oxygen, and forming with it various compounds. This class may be divided as follows:

DIVISION 1.

HYDROGEN, *forming* water.

It has been questioned by some eminent chemists, whether these two last agents should not be classed among the

DIVISION 2.

Bodies forming acids.

NITROGEN, ...*forming* nitric acid.
SULPHUR, ...*forming* sulphuric acid.
PHOSPHORUS, ..*forming* phosphoric acid.
CARBON, ...*forming* carbonic acid.
BORACIUM, ...*forming* boracic acid.
FLUORIUM, ...*forming* fluoric acid.
MURIATIUM, ..*forming* muriatic acid.

DIVISION 3.

Metallic bodies forming alkalie

POTASSIUM, ...*forming* potash.
SODIUM, ...*forming* soda.
AMMONIUM, ...*forming* ammonia.

DIVISION 4.

Metallic bodies forming earths.

CALCIUM, *or metal forming* lime.
MAGNIUM, ...*forming* magnesia.
BARIUM, ...*forming* barytes.
STRONTIUM, ...*forming* strontites.
SILICIUM, ...*forming* silex.
ALUMIUM, ...*forming* alumine.
YTTRIUM, ...*forming* yttria.

---

inflammable bodies, as they are capable of combining with oxygen, as well as with inflammable bodies. But they seem to be more distinctly characterised by their property of supporting combustion than by any other quality.

GLUCIUM, ...*forming* glucina.
ZIRCONIUM, ...*forming* zirconi. *

DIVISION 5.

Metals, either naturally metallic, or yielding their oxygen to carbon or to heat alone.

Subdivision 1.

Malleable Metals.

GOLD,
PLATINA,
PALLADIUM,
SILVER, †
MERCURY, ‡
TIN,
COPPER,
IRON,
LEAD,
NICKEL,
ZINC.

Subdiv. 2.

Brittle Metals.

ARSENIC,
BISMUTH,
ANTIMONY,
MANGANESE,

* Of all these earths, three or four only have as yet been distinctly decomposed.

† These first four metals have commonly been distinguished by the appellation of *perfect* or *noble* metals, on account of their possessing the characteristic properties of ductility, malleability, inalterability, and great specific gravity, in an eminent degree.

‡ Mercury, in its liquid state, cannot, of course, be called a malleable metal. But when frozen, it possesses a considerable degree of malleability.

TELLURIUM,	URANIUM,
COBALT,	COLUMBIUM *or* TAN-
TUNGSTEN,	TALIUM,
MOLYBDENUM,	IRIDIUM,
TITANIUM,	OSMIUM,
CHROME,	RHODIUM. *

CAROLINE.

Oh, what a formidable list! You will have much to do to explain it, Mrs. B.; for I assure you it is perfectly unintelligible to me, and I think rather perplexes than assists me.

MRS. B.

Do not let that alarm you, my dear; I hope that hereafter this classification will appear quite clear, and, so far from perplexing you, will assist you in arranging your ideas. It would be in vain to attempt forming a division that would appear perfectly clear to a beginner: for you may easily conceive that a chemical division being necessarily founded on properties with which you are almost wholly unacquainted, it is impossible that you should at once be able to understand its meaning or appreciate its utility.

* These last four or five metallic bodies are placed under this class for the sake of arrangement, though some of their properties have not been yet fully investigated.

But, before we proceed further, it will be necessary to give you some idea of chemical attraction, a power on which the whole science depends.

Chemical Attraction, or the *Attraction of Composition*, consists in the peculiar tendency which bodies of a different nature have to unite with each other. It is by this force that all the compositions, and decompositions, are effected.

EMILY.

What is the difference between chemical attraction, and the attraction of cohesion, or of aggregation, which you often mentioned to us, in former conversations?

MRS. B.

The attraction of cohesion exists only between particles of the *same* nature, whether simple or compound; thus it unites the particles of a piece of metal which is a simple substance, and likewise the particles of a loaf of bread which is a compound. The attraction of composition, on the contrary, unites and maintains, in a state of combination, particles of a *dissimilar* nature; it is this power that forms each of the compound particles of which bread consists; and it is by the attraction of cohesion that all these particles are connected into a single mass.

EMILY.

The attraction of cohesion, then, is the power which unites the integrant particles of a body: the attraction of composition that which combines the constituent particles. Is it not so?

MRS. B.

Precisely: and observe that the attraction of cohesion unites particles of a similar nature, without changing their original properties; the result of such an union, therefore, is a body of the same kind as the particles of which it is formed; whilst the attraction of composition, by combining particles of a dissimilar nature, produces compound bodies, quite different from any of their constituents. If, for instance, I pour on the piece of copper, contained in this glass, some of this liquid (which is called nitric acid), for which it has a strong attraction, every particle of the copper will combine with a particle of acid, and together they will form a new body, totally different from either the copper or the acid.

Do you observe the internal commotion that already begins to take place? It is produced by the combination of these two substances; and yet the acid has in this case to overcome not only the resistance which the strong cohesion of the particles of copper opposes to their combination with it, but also to overcome the weight of the copper, which makes it sink to the bottom of the glass, and prevents the acid from having such free access to it as it would if the metal were suspended in the liquid.

EMILY.

The acid seems, however, to overcome both these obstacles without difficulty, and appears to be very rapidly dissolving the copper.

MRS. B.

By this means it reduces the copper into more minute parts than could possibly be done by any mechanical power. But as the acid can act only on the surface of the metal, it will be some time before the union of these two bodies will be completed.

You may, however, already see how totally different this compound is from either of its ingredients. It is neither colourless, like the acid, nor hard, heavy, and yellow like the copper. If you tasted it, you would no longer perceive the sourness of the acid. It has at present the appearance of a blue liquid; but when the union is completed, and the water with which the acid is diluted is evaporated, the compound will assume the form of regular crystals, of a fine blue colour, and perfectly transparent *. Of these I can shew you a

* These crystals are more easily obtained from a mixture of sulphuric with a little nitric acid.

specimen, as I have prepared some for that purpose.

CAROLINE.

How very beautiful they are, in colour, form, and transparency!

EMILY.

Nothing can be more striking than this example of chemical attraction.

MRS. B.

The term *attraction* has been lately introduced into chemistry as a substitute for the word *affinity*, to which some chemists have objected, because it originated in the vague notion that chemical combinations depended upon a certain resemblance, or relationship, between particles that are disposed to unite; and this idea is not only imperfect, but erroneous, as it is generally particles of the most dissimilar nature, that have the greatest tendency to combine.

CAROLINE.

Besides, there seems to be no advantage in using a variety of terms to express the same meaning; on the contrary it creates confusion; and as we are well acquainted with the term Attraction in natural philosophy, we had better adopt it in chemistry likewise.

MRS. B.

If you have a clear idea of the meaning, I shall leave you at liberty to express it in the terms you prefer. For myself, I confess that I think the word Attraction best suited to the general law that unites the integrant particles of bodies; and Affinity better adapted to that which combines the constituent particles, as it may convey an idea of the preference which some bodies have for others, which the term *attraction of composition* does not so well express.

EMILY.

So I think; for though that preference may not result from any relationship, or similitude, between the particles (as you say was once supposed), yet, as it really exists, it ought to be expressed.

MRS. B.

Well, let it be agreed that you may use the terms *affinity*, *chemical attraction*, and *attraction of composition*, indifferently, provided you recollect that they have all the same meaning.

EMILY.

I do not conceive how bodies can be decomposed by chemical attraction. That this power should be the means of composing them, is very obvious; but that it should, at the same time, produce exactly the contrary effect, appears to me very singular.

MRS. B.

To decompose a body is, you know, to separate its constituent parts, which, as we have just observed, cannot be done by mechanical means.

EMILY.

No: because mechanical means separate only the integrant particles; they act merely against the attraction of cohesion, and only divide a compound into smaller parts.

MRS. B.

The decomposition of a body is performed by chemical powers. If you present to a body composed of two principles, a third, which has a greater affinity for one of them than the two first have for each other, it will be decomposed, that is, its two principles will be separated by means of the third body. Let us call two ingredients, of which the body is composed, A and B. If we present to it another ingredient C, which has a greater affinity for B than that which unites A and B, it necessarily follows that B will quit A to combine with C. The new ingredient, therefore, has effected a decomposition of the original body A B; A has been left alone, and a new compound, B C, has been formed.

EMILY.

We might, I think, use the comparison of two

friends, who were very happy in each other's society, till a third disunited them by the preference which one of them gave to the new-comer.

MRS. B.

Very well. I shall now show you how this takes place in chemistry.

Let us suppose that we wish to decompose the compound we have just formed by the combination of the two ingredients, copper and nitric acid; we may do this by presenting to it a piece of iron, for which the acid has a stronger attraction than for copper; the acid will, consequently, quit the copper to combine with the iron, and the copper will be what the chemists call *precipitated*, that is to say, it will be thrown down in its separate state, and reappear in its simple form.

In order to produce this effect, I shall dip the blade of this knife into the fluid, and, when I take it out, you will observe, that, instead of being wetted with a bluish liquid, like that contained in the glass, it will be covered with a thin coat of copper.

CAROLINE.

So it is really! but then is it not the copper, instead of the acid, that has combined with the iron blade?

MRS. B.

No; you are deceived by appearances: it is

the acid which combines with the iron, and, in so doing, deposits or precipitates the copper on the surface of the blade.

EMILY.

But, cannot three or more substances combine together, without any of them being precipitated?

MRS. B.

That is sometimes the case; but, in general, the stronger affinity destroys the weaker; and it seldom happens that the attraction of several substances for each other is so equally balanced as to produce such complicated compounds.

CAROLINE.

But, pray, Mrs. B., what is the cause of the chemical attraction of bodies for each other? It appears to me more extraordinary or unnatural, if I may use the expression, than the attraction of cohesion, which unites particles of a similar nature.

MRS. B.

Chemical attraction may, like that of cohesion or gravitation, be one of the powers inherent in matter which, in our present state of knowledge, admits of no other satisfactory explanation than an immediate reference to a divine cause. Sir H. Davy, however, whose important discoveries have opened such improved views in chemistry, has suggested an hypothesis which may throw great light upon that science. He supposes that there are two kinds of electricity, with one or other of which all bodies are united. These we distinguish by the names of *positive* and *negative* electricity; those bodies are disposed to combine, which possess opposite electricities, as they are brought together by the attraction which these electricities have for each other. But, whether this hypothesis be altogether founded on truth or not, it is impossible to question the great influence of electricity in chemical combinations.

EMILY.

So, that we must suppose that the two electricities always attract each other, and thus compel the bodies in which they exist to combine?

CAROLINE.

And may not this be also the cause of the attraction of cohesion?

MRS. B.

No, for in particles of the same nature the same electricities must prevail, and it is only the different or opposite electric fluids that attract each other.

CAROLINE.

These electricities seem to me to be a kind of

chemical spirit, which animates the particles of bodies, and draws them together.

EMILY.

If it is known, then, with which of the electricities bodies are united, it can be inferred which will, and which will not, combine together?

MRS. B.

Certainly. — I should not omit to mention, that some doubts have been entertained whether electricity be really a material agent, or whether it might not be a power inherent in bodies, similar to, or, perhaps identical with, attraction.

EMILY.

But what then would be the electric spark which is visible, aud must therefore be really material?

MRS. B.

What we call the electric spark, may, Sir H. Davy says, be merely the heat and light, or fire produced by the chemical combinations with which these phenomena are always connected. We will not, however, enter more fully on this important subject at present, but reserve the principal facts which relate to it to a future conversation.

Before we part, however, I must recommend you to fix in your memory the names of the simple bodies, against our next interview.

CONVERSATION II.

ON LIGHT AND HEAT OR CALORIC.

CAROLINE.

We have learned by heart the names of all the simple bodies which you have enumerated, and we are now ready to enter on the examination of each of them successively. You will begin, I suppose, with LIGHT?

MRS. B.

Respecting the nature of light we have little more than conjectures. It is considered by most philosophers as a real substance, immediately emanating from the sun, and from all luminous bodies, from which it is projected in right lines with prodigious velocity. Light, however, being imponderable, it cannot be confined and examined by itself; and therefore it is to the effects it produces on other bodies, rather than to its immediate nature, that we must direct our attention.

The connection between light and heat is very obvious; indeed, it is such, that it is extremely difficult to examine the one independently of the other.

EMILY.

But, is it possible to separate light from heat; I thought they were only different degrees of the same thing, fire?

MRS. B.

I told you that fire was not now considered as a simple element. Whether light and heat be altogether different agents, or not, I cannot pretend to decide; but, in many cases, light may be separated from heat. The first discovery of this was made by a celebrated Swedish chemist, Scheele. Another very striking illustration of the separation of heat and light was long after pointed out by Dr. Herschell. This philosopher discovered that these two agents were emitted in the rays of the sun, and that heat was less refrangible than light; for, in separating the different coloured rays of light by a prism (as we did some time ago), he found that the greatest heat was beyond the spectrum, at a little distance from the red rays, which, you may recollect, are the least refrangible.

EMILY.

I should like to try that experiment.

MRS. B.

It is by no means an easy one: the heat of a ray of light, refracted by a prism, is so small, that it requires a very delicate thermometer to distinguish the difference of the degree of heat within and without the spectrum. For in this experiment the heat is not totally separated from the light, each coloured ray retaining a certain portion of it, though the greatest part is not sufficiently refracted to fall within the spectrum.

EMILY.

I suppose, then, that those coloured rays which are the least refrangible, retain the greatest quantity of heat?

MRS. B.

They do so.

EMILY.

Though I no longer doubt that light and heat can be separated, Dr. Herschell's experiment does not appear to me to afford sufficient proof that they are essentially different; for light, which you call a simple body, may likewise be divided into the various coloured rays.

MRS. B.

No doubt there must be some difference in the various coloured rays. Even their chemical powers are different. The blue rays, for instance, have the greatest effect in separating oxygen from bodies, as was found by Scheele; and there exist also, as Dr. Wollaston has shown, rays more refrangible than the blue, which produce the same chemical effect, and, what is very remarkable, are invisible.

EMILY.

Do you think it possible that heat may be merely a modification of light?

MRS. B.

That is a supposition which, in the present state of natural philosophy, can neither be positively affirmed nor denied. Let us, therefore, instead of discussing theoretical points, be contented with examining what is known respecting the chemical effects of light.

Light is capable of entering into a kind of transitory union with certain substances, and this is what has been called phosphorescence. Bodies that are possessed of this property, after being exposed to the sun's rays, appear luminous in the dark. The shells of fish, the bones of land animals, marble, limestone, and a variety of combinations of earths, are more or less powerfully phosphorescent.

CAROLINE.

I remember being much surprised last summer with the phosphorescent appearance of some pieces of rotten wood, which had just been dug out of the ground; they shone so bright that I at first supposed them to be glow-worms.

EMILY.

And is not the light of a glow-worm of a phosphorescent nature?

MRS. B.

It is a very remarkable instance of phosphorescence in living animals; this property, however, is not exclusively possessed by the glow-worm. The insect called the lanthorn-fly, which is peculiar to warm climates, emits light as it flies, producing in the dark a remarkably sparkling appearance. But it is more common to see animal matter in a dead state possessed of a phosphorescent quality; sea fish is often eminently so.

EMILY.

I have heard that the sea has sometimes had the appearance of being illuminated, and that the light is supposed to proceed from the spawn of fishes floating on its surface.

MRS. B.

This light is probably owing to that or some other animal matter. Sea water has been observed to become luminous from the substance of a fresh herring having been immersed in it; and certain insects, of the Medusa kind, are known to produce similar effects.

But the strongest phosphorescence is produced by chemical compositions prepared for the purpose, the most common of which consists of oyster shells and sulphur, and is known by the name of Canton's Phosphorus.

EMILY.

I am rather surprised, Mrs. B., that you should have said so much of the light emitted by phosphorescent bodies without taking any notice of that which is produced by burning bodies.

MRS. B.

The light emitted by the latter is so intimately connected with the chemical history of combustion, that I must defer all explanation of it till we come to the examination of that process, which is one of the most interesting in chemical science.

Light is an agent capable of producing various chemical changes. It is essential to the welfare both of the animal and vegetable kingdoms; for men and plants grow pale and sickly if deprived of its salutary influence. It is likewise remarkable for its property of destroying colour, which renders it of great consequence in the process of bleaching.

EMILY.

Is it not singular that light, which in studying optics we were taught to consider as the source and origin of colours, should have also the power of destroying them?

CAROLINE.

It is a fact, however, that we every day experience; you know how it fades the colours of linens and silks.

EMILY.

Certainly. And I recollect that endive is made to grow white instead of green, by being covered up so as to exclude the light. But by what means does light produce these effects?

MRS. B.

This I cannot attempt to explain to you until you have obtained a further knowledge of chemistry. As the chemical properties of light can be accounted for only in their reference to compound bodies, it would be useless to detain you any longer on this subject; we may therefore pass on to the examination of heat, or caloric, with which we are somewhat better acquainted.

HEAT and LIGHT may be always distinguished by the different sensations they produce, *Light* affects the sense of sight; *Caloric* that of feeling; the one produces *Vision*, the other the sensation of *Heat*.

Caloric is found to exist in a variety of forms or modifications, and I think it will be best to consider it under the two following heads, viz.

1. FREE OR RADIANT CALORIC.
2. COMBINED CALORIC.

The first, FREE OR RADIANT CALORIC, is also called HEAT OF TEMPERATURE; it comprehends all heat which is perceptible to the senses, and affects the thermometer.

EMILY.

You mean such as the heat of the sun, of fire, of candles, of stoves; in short, of every thing that burns?

MRS. B.

And likewise of things that do not burn, as, for instance, the warmth of the body; in a word, all heat that is *sensible*, whatever may be its degree, or the source from which it is derived.

CAROLINE.

What then are the other modifications of caloric? It must be a strange kind of heat that cannot be perceived by our senses.

MRS. B.

None of the modifications of caloric should properly be called *heat;* for heat, strictly speaking, is the sensation produced by caloric, on animated bodies; this word, therefore, in the accurate language of science, should be confined to express the sensation. But custom has adapted it likewise to inanimate matter, and we say *the heat of an oven, the heat of the sun,* without any reference to the sensation which they are capable of exciting.

It was in order to avoid the confusion which arose from thus confounding the cause and effect, that modern chemists adopted the new word *caloric,* to denote the principle which produces heat; yet they do not always, in compliance with their own language, limit the word *heat* to the expression of the sensation, since they still frequently employ it in reference to the other modifications of caloric which are quite independent of sensation.

CAROLINE.

But you have not yet explained to us what these other modifications of caloric are.

MRS. B.

Because you are not acquainted with the properties of free caloric, and you know that we have agreed to proceed with regularity.

One of the most remarkable properties of free caloric is its power of *dilating* bodies. This fluid is so extremely subtle, that it enters and pervades all bodies whatever, forces itself between their particles, and not only separates them, but frequently drives them asunder to a considerable distance from each other. It is thus that caloric dilates or expands a body so as to make it occupy a greater space than it did before.

EMILY.

The effect it has on bodies, therefore, is directly contrary to that of the attraction of cohesion; the one draws the particles together, the other drives them asunder.

MRS. B.

Precisely. There is a continual struggle between the attraction of aggregation, and the expansive power of caloric; and from the action of these two opposite forces, result all the various forms of matter, or degrees of consistence, from the solid, to the liquid and aëriform state. And accordingly we find that most bodies are capable of passing from one of these forms to the other, merely in consequence of their receiving different quantities of caloric.

CAROLINE.

That is very curious; but I think I understand the reason of it. If a great quantity of caloric is added to a solid body, it introduces itself between the particles in such a manner as to overcome, in a considerable degree, the attraction of cohesion; and the body, from a solid, is then converted into a fluid.

MRS. B.

This is the case whenever a body is fused or melted; but if you add caloric to a liquid, can you tell me what is the consequence?

CAROLINE.

The caloric forces itself in greater abundance between the particles of the fluid, and drives them to such a distance from each other, that their attraction of aggregation is wholly destroyed: the liquid is then transformed into vapour.

MRS. B.

Very well; and this is precisely the case with boiling water, when it is converted into steam or vapour, and with all bodies that assume an aëriform state.

EMILY.

I do not well understand the word aëriform?

MRS. B.

Any elastic fluid whatever, whether it be merely vapour or permanent air, is called aëriform.

But each of these various states, solid, liquid, and aëriform, admit of many different degrees of density, or consistence, still arising (chiefly at least) from the different quantities of caloric the bodies contain. Solids are of various degrees of density, from that of gold, to that of a thin jelly. Liquids, from the consistence of melted glue, or melted metals, to that of ether, which is the lightest of all liquids. The different elastic fluids (with which you are not yet acquainted) are susceptible of no less variety in their degrees of density.

EMILY.

But does not every individual body also admit of different degrees of consistence, without changing its state?

MRS. B.

Undoubtedly; and this I can immediately show you by a very simple experiment. This piece of iron now exactly fits the frame, or ring, made to receive it; but if heated red hot, it will no longer do so, for its dimensions will be so much increased by the caloric that has penetrated into it, that it will be much too large for the frame.

The iron is now red hot; by applying it to the frame, we shall see how much it is dilated.

EMILY.

Considerably so indeed! I knew that heat had this effect on bodies, but I did not imagine that it could be made so conspicuous.

MRS. B.

By means of this instrument (called a Pyrometer) we may estimate, in the most exact manner, the various dilatations of any solid body by heat. The body we are now going to submit to trial is this small iron bar; I fix it to this apparatus, (PLATE I. Fig I.) and then heat it by lighting the three lamps beneath it: when the bar expands, it increases in length as well as thickness; and, as one end communicates with this wheel-work, whilst the other end is fixed and immoveable, no sooner does it begin to dilate than it presses against the wheel-work, and sets in motion the index, which points out the degrees of dilatation on the dial-plate.

EMILY.

This is, indeed, a very curious instrument; but I do not understand the use of the wheels: would it not be more simple, and answer the purpose equally well, if the bar, in dilating, pressed against the index, and put it in motion without the intervention of the wheels?

Fig. 2.

Fig. 1

Pyrometer.

A A B B B B C A A 1 2 3

Fig. 1 A.A Bar of Metal. 1.2.3 Lamps burning. B.B Wheel work. C Index. Fig. 2 A.A Glass tubes with bulbs. B.B Glasses of water in which they are immers

Drawn by the Author.

Published by Longman & Co. Octr 2nd 1809.

Engraved by Lowry

MRS. B.

The use of the wheels is merely to multiply the motion, and therefore render the effect of the caloric more obvious; for if the index moved no more than the bar increased in length, its motion would scarcely be perceptible; but by means of the wheels it moves in a much greater proportion, which therefore renders the variations far more conspicuous.

By submitting different bodies to the test of the pyrometer, it is found that they are far from dilating in the same proportion. Different metals expand in different degrees, and other kinds of solid bodies vary still more in this respect. But this different susceptibility of dilatation is still more remarkable in fluids than in solid bodies, as I shall show you. I have here two glass tubes, terminated at one end by large bulbs. We shall fill the bulbs, the one with spirit of wine, the other with water. I have coloured both liquids, in order that the effect may be more conspicuous. The spirit of wine, you see, dilates by the warmth of my hand as I hold the bulb.

EMILY.

It certainly does, for I see it is rising into the tube. But water, it seems, is not so easily affected by heat; for scarcely any change is produced on it by the warmth of the hand.

MRS. B.

True; we shall now plunge the bulbs into hot water, (PLATE I. Fig. 2.) and you will see both liquids rise in the tubes; but the spirit of wine will ascend highest.

CAROLINE.

How rapidly it expands! Now it has nearly reached the top of the tube, though the water has hardly begun to rise.

EMILY.

The water now begins to dilate. Are not these glass tubes, with liquids rising within them, very like thermometers?

MRS. B.

A thermometer is constructed exactly on the same principle, and these tubes require only a scale to answer the purpose of thermometers: but they would be rather awkward in their dimensions. The tubes and bulbs of thermometers, though of various sizes, are in general much smaller than these; the tube too is hermetically closed, and the air excluded from it. The fluid most generally used in thermometers is mercury, commonly called quicksilver, the dilatations and contractions of which correspond more exactly to the additions, and subtractions, of caloric, than those of any other fluid.

CAROLINE.

Yet I have often seen coloured spirit of wine used in thermometers.

MRS. B.

The expansions and contractions of that liquid are not quite so uniform as those of mercury; but in cases in which it is not requisite to ascertain the temperature with great precision, spirit of wine will answer the purpose equally well, and indeed in some respects better, as the expansion of the latter is greater, and therefore more conspicuous. This fluid is used likewise in situations and experiments in which mercury would be frozen; for mercury becomes a solid body, like a piece of lead or any other metal, at a certain degree of cold: but no degree of cold has ever been known to freeze spirit of wine.

A thermometer, therefore, consists of a tube with a bulb, such as you see here, containing a fluid whose degrees of dilatation and contraction are indicated by a scale to which the tube is fixed. The degree which indicates the boiling point, simply means that, when the fluid is sufficiently dilated to rise to this point, the heat is such that water exposed to the same temperature will boil. When, on the other hand, the fluid is so much condensed as to sink to the freezing point, we know that water will freeze at that temperature. The extreme points of the scales are not the same in all thermometers, nor are the degrees always divided in the same manner. In different countries philosophers have chosen to adopt different scales and divisions. The two thermometers most used are those of Fahrenheit, and of Reaumur; the first is generally preferred by the English, the latter by the French.

EMILY.

The variety of scale must be very inconvenient, and I should think liable to occasion confusion, when French and English experiments are compared.

MRS. B.

The inconvenience is but very trifling, because the different gradations of the scales do not affect the principle upon which thermometers are constructed. When we know, for instance, that Fahrenheit's scale is divided into 212 degrees, in which 32° corresponds with the freezing point, and 212° with the point of boiling water: and that Reaumur's is divided only into 80 degrees, in which 0° denotes the freezing point, and 80° that of boiling water, it is easy to compare the two scales together, and reduce the one into the other. But, for greater convenience, thermometers are sometimes constructed with both these scales, one

THERMOMETER.

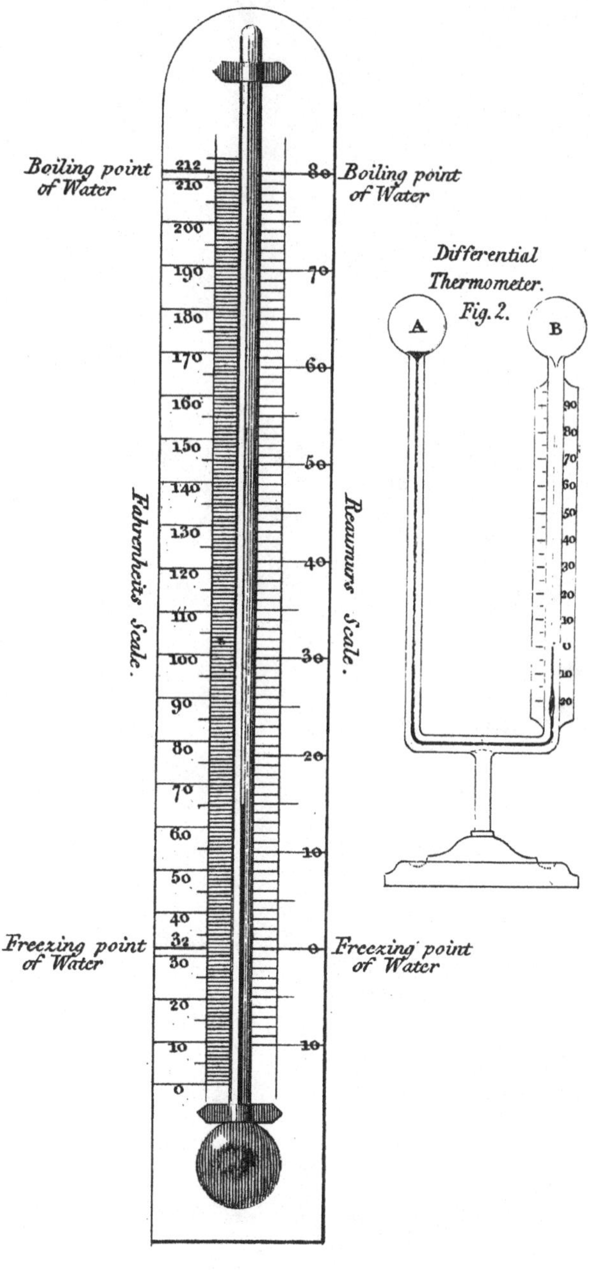

Drawn by the Author. Published by Longman & C°. Oct.r 2nd 1809. Engraved by Lowry.

on either side of the tube; so that the correspondence of the different degrees of the two scales is thus instantly seen. Here is one of these scales, (Plate II. Fig. 1.) by which you can at once perceive that each degree of Reaumur's corresponds to 2¼ of Fahrenheit's division. But I believe the French have, of late, given the preference to what they call the centigrade scale, in which the space between the freezing and the boiling point is divided into 100 degrees.

CAROLINE.

That seems to me the most reasonable division, and I cannot guess why the freezing point is called 32°, or what advantage is derived from it.

MRS. B.

There really is no advantage in it; and it originated in a mistaken opinion of the instrument-maker, Fahrenheit, who first constructed these thermometers. He mixed snow and salt together, and produced by that means a degree of cold which he concluded was the greatest possible, and therefore made his scale begin from that point. Between that and boiling water he made 212 degrees, and the freezing point was found to be at 32°.

EMILY.

Are spirit of wine, and mercury, the only liquids used in the construction of thermometers?

MRS. B.

I believe they are the only liquids now in use, though some others, such as linseed oil, would make tolerable thermometers: but for experiments in which a very quick and delicate test of the changes of temperature is required, air is the fluid sometimes employed. The bulb of air thermometers is filled with common air only, and its expansion and contraction are indicated by a small drop of any coloured liquor, which is suspended within the tube, and moves up and down, according as the air within the bulb and tube expands or contracts. But in general, air thermometers, however sensible to changes of temperature, are by no means accurate in their indications.

I can, however, show you an air thermometer of a very peculiar construction, which is remarkably well adapted for some chemical experiments, as it is equally delicate and accurate in its indications.

CAROLINE.

It looks like a double thermometer reversed, the tube being bent, and having a large bulb at each of its extremities. (Plate II. Fig. 2.)

EMILY.

Why do you call it an air thermometer; the tube contains a coloured liquid?

MRS. B.

But observe that the bulbs are filled with air, the liquid being confined to a portion of the tube, and answering only the purpose of showing, by its motion in the tube, the comparative dilatation or contraction of the air within the bulbs, which afford an indication of their relative temperature. Thus if you heat the bulb A, by the warmth of your hand, the fluid will rise towards the bulb B, and the contrary will happen if you reverse the experiment.

But if, on the contrary, both tubes are of the same temperature, as is the case now, the coloured liquid, suffering an equal pressure on each side, no change of level takes place.

CAROLINE.

This instrument appears, indeed, uncommonly delicate. The fluid is set in motion by the mere approach of my hand.

MRS. B.

You must observe, however, that this thermometer cannot indicate the temperature of any particular body, or of the medium in which it is immersed; it serves only to point out the *difference* of temperature between the two bulbs, when placed under different circumstances. For this reason it has been called *differential* thermometer. You will see by-and-bye to what particular purposes this instrument applies.

EMILY.

But do common thermometers indicate the exact quantity of caloric contained either in the atmosphere, or in any body with which they are in contact?

MRS. B.

No: first, because there are other modifications of caloric which do not affect the thermometer; and, secondly, because the temperature of a body, as indicated by the thermometer, is only relative. When, for instance, the thermometer remains stationary at the freezing point, we know that the atmosphere (or medium in which it is placed, whatever it may be) is as cold as freezing water; and when it stands at the boiling point, we know that this medium is as hot as boiling water; but we do not know the positive quantity of heat contained either in freezing or boiling water, any more than we know the real extremes of heat and cold; and consequently we cannot determine that of the body in which the thermometer is placed.

CAROLINE.

I do not quite understand this explanation.

MRS. B.

Let us compare a thermometer to a well, in which the water rises to different heights, according as it is more or less supplied by the spring which feeds it: if the depth of the well is unfathomable, it must be impossible to know the absolute quantity of water it contains; yet we can with the greatest accuracy measure the number of feet the water has risen or fallen in the well at any time, and consequently know the precise quantity of its increase or diminution, without having the least knowledge of the whole quantity of water it contains.

CAROLINE.

Now I comprehend it very well; nothing appears to me to explain a thing so clearly as a comparison.

EMILY.

But will thermometers bear any degree of heat?

MRS. B.

No; for if the temperature were much above the highest degree marked on the scale of the thermometer, the mercury would burst the tube in an attempt to ascend. And at any rate, no thermometer can be applied to temperatures higher than the boiling point of the liquid used in its construction, for the steam, on the liquid beginning to boil, would burst the tube. In furnaces, or whenever any very high temperature is to be measured, a pyrometer, invented by Wedgwood, is used for that purpose. It is made of a certain composition of baked clay, which has the peculiar property of contracting by heat, so that the degree of contraction of this substance indicates the temperature to which it has been exposed.

EMILY.

But is it possible for a body to contract by heat? I thought that heat dilated all bodies whatever.

MRS. B.

This is not an exception to the rule. You must recollect that the bulk of the clay is not compared, whilst hot, with that which it has when cold; but it is from the change which the clay has undergone by *having been* heated that the indications of this instrument are derived. This change consists in a beginning fusion which tends to unite the particles of clay more closely, thus rendering it less pervious or spongy.

Clay is to be considered as a spongy body, having many interstices or pores, from its having contained water when soft. These interstices are by heat lessened, and would by extreme heat be entirely obliterated.

CAROLINE.

And how do you ascertain the degrees of contraction of Wedgwood's pyrometer?

MRS. B.

The dimensions of a piece of clay are measured by a scale graduated on the side of a tapered groove, formed in a brass ruler; the more the clay is contracted by the heat, the further it will descend into the narrow part of the tube.

Before we quit the subject of expansion, I must observe to you that, as liquids expand more readily than solids, so elastic fluids, whether air or vapour, are the most expansible of all bodies.

It may appear extraordinary that all elastic fluids whatever, undergo the same degree of expansion from equal augmentations of temperature.

EMILY.

I suppose, then, that all elastic fluids are of the same density?

MRS. B.

Very far from it; they vary in density, more than either liquids or solids. The uniformity of their expansibility, which at first may appear singular, is, however, readily accounted for. For if the different susceptibilities of expansion of bodies arise from their various degrees of attraction of cohesion, no such difference can be expected in elastic fluids, since in these the attraction of cohesion does not exist, their particles being on the contrary possessed of an elastic or repulsive power; they will therefore all be equally expanded by equal degrees of caloric.

EMILY.

True; as there is no power opposed to the expansive force of caloric in elastic bodies, its effect must be the same in all of them.

MRS. B.

Let us now proceed to examine the other properties of free caloric.

Free caloric always tends to diffuse itself equally, that is to say, when two bodies are of different temperatures, the warmer gradually parts with its heat to the colder, till they are both brought to the same temperature. Thus, when a thermometer is applied to a hot body, it receives caloric; when to a cold one, it communicates part of its own caloric, and this communication continues until the thermometer and the body arrive at the same temperature.

EMILY.

Cold, then, is nothing but a negative quality, simply implying the absence of heat.

MRS. B.

Not the total absence, but a diminution of heat; for we know of no body in which some caloric may not be discovered.

CAROLINE.

But when I lay my hand on this marble table I feel it *positively* cold, and cannot conceive that there is any caloric in it.

MRS. B.

The cold you experience consists in the loss of caloric that your hand sustains in an attempt to bring its temperature to an equilibrium with the marble. If you lay a piece of ice upon it, you will find that the contrary effect will take place; the ice will be melted by the heat which it abstracts from the marble.

CAROLINE.

Is it not in this case the air of the room, which being warmer than the marble, melts the ice?

MRS. B.

The air certainly acts on the surface which is exposed to it, but the table melts that part with which it is in contact.

CAROLINE.

But why does caloric tend to an equilibrium?

It cannot be on the same principle as other fluids, since it has no weight?

MRS. B.

Very true, Caroline, that is an excellent objection. You might also, with some propriety, object to the term *equilibrium* being applied to a body that is without weight; but I know of no expression that would explain my meaning so well. You must consider it, however, in a figurative rather than a literal sense; its strict meaning is an *equal diffusion*. We cannot, indeed, well say by what power it diffuses itself equally, though it is not surprising that it should go from the parts which have the most to those which have the least. This subject is best explained by a theory suggested by Professor Prevost of Geneva, which is now, I believe, generally adopted.

According to this theory, caloric is composed of particles perfectly separate from each other, every one of which moves with a rapid velocity in a certain direction. These directions vary as much as imagination can conceive, the result of which is, that there are rays or lines of these particles moving with immense velocity in every possible direction. Caloric is thus universally diffused, so that when any portion of space happens to be in the neighbourhood of another, which contains more caloric, the colder portion receives a quantity of calorific rays from the latter, sufficient to restore an equilibrium of temperature. This radiation does not only take place in free space, but extends also to bodies of every kind. Thus you may suppose all bodies whatever constantly radiating caloric: those that are of the same temperature give out and absorb equal quantities, so that no variation of temperature is produced in them; but when one body contains more free caloric than another, the exchange is always in favour of the colder body, until an equilibrium is effected; this you found to be the case when the marble table cooled your hand, and again when it melted the ice.

CAROLINE.

This reciprocal radiation surprises me extremely; I thought, from what you first said, that the hotter bodies alone emitted rays of caloric which were absorbed by the colder; for it seems unnatural that a hot body should receive any caloric from a cold one, even though it should return a greater quantity.

MRS. B.

It may at first appear so, but it is no more extraordinary than that a candle should send forth rays of light to the sun, which, you know, must necessarily happen.

CAROLINE.

Well, Mrs. B—, I believe that I must give up the point. But I wish I could *see* these rays of caloric; I should then have greater faith in them.

MRS. B.

Will you give no credit to any sense but that of sight? You may feel the rays of caloric which you receive from any body of a temperature higher than your own; the loss of the caloric you part with in return, it is true, is not perceptible; for as you gain more than you lose, instead of suffering a diminution, you are really making an acquisition of caloric. It is, therefore, only when you are parting with it to a body of a lower temperature, that you are sensible of the sensation of cold, because you then sustain an absolute loss of caloric.

EMILY.

And in this case we cannot be sensible of the small quantity of heat we receive in exchange from the colder body, because it serves only to diminish the loss.

MRS. B.

Very well, indeed, Emily. Professor Pictet, of Geneva, has made some very interesting experiments, which prove not only that caloric radiates from all bodies whatever, but that these rays may be reflected, according to the laws of optics, in

MR. PICTET'S APPARATUS FOR THE REFLECTION OF HEAT.

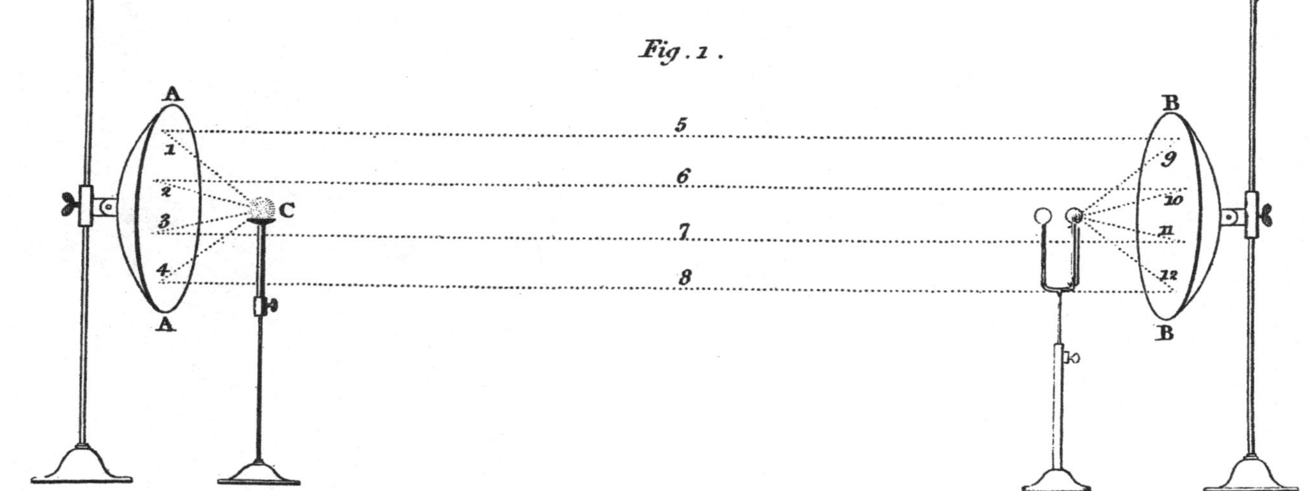

A.A & B.B Concave mirrors fixed on stands.—C Heated Bullet placed in the focus of the mirror A.—D Thermometer, with its bulb placed in the focus of the mirror B.—1. 2. 3. 4 Rays of Caloric radiating from the bullet & falling on the mirror A.—5. 6. 7. 8 The same rays reflected from the mirror A to the mirror B.—9. 10. 11. 12 The same rays reflected by the mirror B to the Thermometer.

Drawn by the Author.

Engraved by Lowry.

Published by Longman & Co. Octr. 2nd. 1809.

the same manner as light. I shall repeat these experiments before you, having procured mirrors fit for the purpose; and it will afford us an opportunity of using the differential thermometer, which is particularly well adapted for these experiments. — I place an iron bullet, (Plate III. Fig. I.) about two inches in diameter, and heated to a degree not sufficient to render it luminous, in the focus of this large metallic concave mirror. The rays of heat which fall on this mirror are reflected, agreeably to the property of concave mirrors, in a parallel direction, so as to fall on a similar mirror, which, you see, is placed opposite to the first, at the distance of about ten feet; thence the rays converge to the focus of the second mirror, in which I place one of the bulbs of this thermometer. Now, observe in what manner it is affected by the caloric which is reflected on it from the heated bullet. — The air is dilated in the bulb which we placed in the focus of the mirror, and the liquor rises considerably in the opposite leg.

EMILY.

But would not the same effect take place, if the rays of caloric from the heated bullet fell directly on the thermometer, without the assistance of the mirrors?

MRS. B.

The effect would in that case be so trifling, at

the distance at which the bullet and the thermometer are from each other, that it would be almost imperceptible. The mirrors, you know, greatly increase the effect, by collecting a large quantity of rays into a focus; place your hand in the focus of the mirror, and you will find it much hotter there than when you remove it nearer to the bullet.

EMILY.

That is very true; it appears extremely singular to feel the heat diminish in approaching the body from which it proceeds.

CAROLINE.

And the mirror which produces so much heat, by converging the rays, is itself quite cold.

MRS. B.

The same number of rays that are dispersed over the surface of the mirror are collected by it into the focus; but, if you consider how large a surface the mirror presents to the rays, and, consequently, how much they are diffused in comparison to what they are at the focus, which is little more than a point, I think you can no longer wonder that the focus should be so much hotter than the mirror.

The principal use of the mirrors in this experiment is, to prove that the caloric emanation is reflected in the same manner as light.

CAROLINE.

And the result, I think, is very conclusive.

MRS. B.

The experiment may be repeated with a wax taper instead of the bullet, with a view of separating the light from the caloric. For this purpose a transparent plate of glass must be interposed between the mirrors; for light, you know, passes with great facility through glass, whilst the transmission of caloric is almost wholly impeded by it. We shall find, however, in this experiment, that some few of the calorific rays pass through the glass together with the light, as the thermometer rises a little; but, as soon as the glass is removed, and a free passage left to the caloric, it will rise considerably higher.

EMILY.

This experiment, as well as that of Dr. Herschell's, proves that light and heat may be separated; for in the latter experiment the separation was not perfect, any more than in that of Mr. Pictet.

CAROLINE.

I should like to repeat this experiment, with the difference of substituting a cold body instead of the hot one, to see whether cold would not be reflected as well as heat.

MRS. B.

That experiment was proposed to Mr. Pictet by an incredulous philosopher like yourself, and he immediately tried it by substituting a piece of ice in the place of the heated bullet.

CAROLINE.

Well, Mrs. B., and what was the result?

MRS. B.

That we shall see; I have procured some ice for the purpose.

EMILY.

The thermometer falls considerably

CAROLINE.

And does not that prove that cold is not merely a *negative* quality, implying simply an inferior degree of heat? The cold must be *positive*, since it is capable of reflection.

MRS. B.

So it at first appeared to Mr. Pictet; but upon a little consideration he found that it afforded only an additional proof of the reflection of heat: this I shall endeavour to explain to you.

According to Mr. Prevost's theory, we suppose that all bodies whatever radiate caloric; the thermometer used in these experiments therefore emits calorific rays in the same manner as any other substance. When its temperature is in equilibrium with that of the surrounding bodies, it receives as much caloric as it parts with, and no change of temperature is produced. But when we introduce a body of a lower temperature, such as a piece of ice, which parts with less caloric than it receives, the consequence is, that its temperature is raised, whilst that of the surrounding bodies is proportionally lowered.

EMILY.

If, for instance, I was to bring a large piece of ice into this room, the ice would in time be melted, by absorbing caloric from the general radiation which is going on throughout the room; and as it would contribute very little caloric in return for what is absorbed, the room would necessarily be cooled by it.

MRS. B.

Just so; and as in consequence of the mirrors, a more considerable exchange of rays takes place between the ice and the thermometer, than between these and any of the surrounding bodies, the temperature of the thermometer must be more lowered than that of any other adjacent object.

CAROLINE.

I confess I do not perfectly understand your explanation.

MRS. B.

This experiment is exactly similar to that made with the heated bullet: for, if we consider the thermometer as the hot body (which it certainly is in comparison to the ice), you may then easily understand that it is by the loss of the calorific rays which the thermometer sends to the ice, and not by any cold rays received from it, that the fall of the mercury is occasioned: for the ice, far from emitting rays of cold, sends forth rays of caloric, which diminish the loss sustained by the thermometer.

Let us say, for instance, that the radiation of the thermometer towards the ice is equal to 20, and that of the ice towards the thermometer to 10: the exchange in favour of the ice is as 20 is to 10, or the thermometer absolutely loses 10, whilst the ice gains 10.

CAROLINE.

But if the ice actually sends rays of caloric to the thermometer, must not the latter fall still lower when the ice is removed?

MRS. B.

No; for the space that the ice occupied, admits rays from all the surrounding bodies to pass through it; and those being of the same temperature as the thermometer, will not affect it, because as much heat now returns to the thermometer as radiates from it.

CAROLINE.

I must confess that you have explained this in so satisfactory a manner, that I cannot help being convinced now that cold has no real claim to the rank of a positive being.

MRS. B.

Before I conclude the subject of radiation I must observe to you that different bodies, (or rather surfaces,) possess the power of radiating caloric in very different degrees.

Some very curious experiments have been made by Mr. Leslie on this subject, and it was for this purpose that he invented the differential thermometer; with its assistance he ascertained that black surfaces radiate most, glass next, and polished surfaces the least of all.

EMILY.

Supposing these surfaces, of course, to be all of the same temperature.

MRS. B.

Undoubtedly. I will now show you the very simple and ingenious apparatus, by means of which he made these experiments. This cubical tin vessel or canister, has each of its sides externally covered with different materials; the one is simply blackened; the next is covered with white paper; the third with a pane of glass, and in the fourth the polished tin surface remains uncovered. We shall fill this vessel with hot water, so that there can be no doubt but that all its sides will be of the same temperature. Now let us place it in the focus of one of the mirrors, making each of its sides front it in succession. We shall begin with the black surface.

CAROLINE.

It makes the thermometer which is in the focus of the other mirror rise considerably.—Let us turn the paper surface towards the mirror. The thermometer falls a little, therefore of course this side cannot emit or radiate so much caloric as the blackened side.

EMILY.

This is very surprising; for the sides are exactly of the same size, and must be of the same temperature. But let us try the glass surface.

MRS. B.

The thermometer continues falling, and with the plain surface it falls still lower; these two surfaces therefore radiate less and less.

CAROLINE.

I think I have found out the reason of this.

MRS. B.

I should be very happy to hear it, for it has not yet (to my knowledge) been accounted for.

CAROLINE.

The water within the vessel gradually cools, and the thermometer in consequence gradually falls.

MRS. B.

It is true that the water cools, but certainly in much less proportion than the thermometer descends, as you will perceive if you now change the tin surface for the black one.

CAROLINE.

I was mistaken certainly, for the thermometer rises again now that the black surface fronts the mirror.

MRS. B.

And yet the water in the vessel is still cooling, Caroline.

EMILY.

I am surprised that the tin surface should radiate the least carolic, for a metallic vessel filled with hot water, a silver teapot, for instance, feels much hotter to the hand than one of black earthen ware.

MRS. B.

That is owing to the different power which various bodies possess for *conducting* caloric, a property which we shall presently examine. Thus, although a metallic vessel feels warmer to the hand, a vessel of this kind is known to preserve the heat of the liquid within, better than one of any other materials; it is for this reason that silver teapots make better tea than those of earthen ware.

EMILY.

According to these experiments, light-coloured dresses, in cold weather, should keep us warmer than black clothes, since the latter radiate so much more than the former.

MRS. B.

And that is actually the case.

EMILY.

This property, of different surfaces to radiate in different degrees, appears to me to be at variance with the equilibrium of caloric; since it would imply that those bodies which radiate most, must ultimately become coldest.

Suppose that we were to vary this experiment, by using two metallic vessels full of boiling water, the one blackened, the other not; would not the black one cool the first?

CAROLINE.

True; but when they were both brought down to the temperature of the room, the interchange of caloric between the canisters and the other bodies of the room being then equal, their temperatures would remain the same.

EMILY.

I do not see why that should be the case; for if different surfaces of the same temperature radiate in different degrees when heated, why should they not continue to do so when cooled down to the temperature of the room?

MRS. B.

You have started a difficulty, Emily, which certainly requires explanation. It is found by experiment that the power of absorption corresponds with and is proportional to that of radiation; so that under equal temperatures, bodies compensate for the greater loss they sustain in consequence of their greater radiation by their greater absorption; so that if you were to make your experiment in an atmosphere heated like the canisters, to the temperature of boiling water, though it is true that the canisters would radiate in different degrees, no change of temperature would be produced in them, because they would each absorb caloric in proportion to their respective radiation.

EMILY.

But would not the canisters of boiling water also absorb caloric in different degrees in a room of the common temperature?

MRS. B.

Undoubtedly they would. But the various bodies in the room would not, at a lower temperature, furnish either of the canisters with a sufficiency of caloric to compensate for the loss they undergo; for, suppose the black canister to absorb 400 rays of caloric, whilst the metallic one absorbed only 200; yet if the former radiate 800, whilst the latter radiates only 400, the black canister will be the first cooled down to the temperature of the room. But from the moment the equilibrium of temperature has taken place, the black canister, both receiving and giving out 400 rays, and the metallic one 200, no change of temperature will take place.

EMILY.

I now understand it extremely well. But what becomes of the surplus of calorific rays, which good radiators emit and bad radiators refuse to receive; they must wander about in search of a resting-place?

MRS. B.

They really do so; for they are rejected and sent

back, or, in other words, *reflected* by the bodies which are bad radiators of caloric; and they are thus transmitted to other bodies which happen to lie in their way, by which they are either absorbed or again reflected, according as the property of reflection, or that of absorption, predominates in these bodies.

CAROLINE.

I do not well understand the difference between radiating and reflecting caloric, for the caloric that is reflected from a body proceeds from it in straight lincs, and may surely be said to radiate from it?

MRS. B.

It is true that there at first appears to be a great analogy between *radiation* and *reflection*, as they equally convey the idea of the transmission of caloric.

But if you consider a little, you will perceive that when a body *radiates* caloric, the heat which it emits not only proceeds from, but has its origin in the body itself. Whilst when a body *reflects* caloric, it parts with none of its own caloric, but only reflects that which it receives from other bodies.

EMILY.

Of this difference we have very striking examples before us, in the tin vessel of water, and the concave mirrors; the first radiates its own heat, the latter reflect the heat which they receive from other bodies.

CAROLINE.

Now, that I understand the difference, it no longer surprises me that bodies which radiate, or part with their own caloric freely, should not have the power of transmitting with equal facility that which they receive from other bodies.

EMILY.

Yet no body can be said to possess caloric of its own, if all caloric is originally derived from the sun.

MRS. B.

When I speak of a body radiating its own caloric, I mean that which it has absorbed and incorporated either immediately from the sun's rays, or through the medium of any other substance.

CAROLINE.

It seems natural enough that the power of absorption should be in opposition to that of reflection, for the more caloric a body receives, the less it will reject.

EMILY.

And equally so that the power of radiation should correspond with that of absorption. It is, in fact, cause and effect; for a body cannot radiate heat without having previously absorbed it; just as a spring that is well fed flows abundantly.

MRS. B.

Fluids are in general very bad radiators of caloric; and air neither radiates nor absorbs caloric in any sensible degree.

We have not yet concluded our observations on free caloric. But I shall defer, till our next meeting, what I have further to say on this subject. I believe it will afford us ample conversation for another interview.

CONVERSATION III.

CONTINUATION OF THE SUBJECT.

MRS. B.

In our last conversation, we began to examine the tendency of caloric to restore an equilibrium of temperature. This property, when once well understood, affords the explanation of a great variety of facts which appeared formerly unaccountable. You must observe, in the first place, that the effect of this tendency is gradually to bring all bodies that are in contact to the same temperature. Thus, the fire which burns in the grate, communicates its heat from one object to another, till every part of the room has an equal proportion of it.

EMILY.

And yet this book is not so cold as the table on which it lies, though both are at an equal distance from the fire, and actually in contact with each other, so that, according to your theory, they should be exactly of the same temperature.

CAROLINE.

And the hearth, which is much nearer the fire than the carpet, is certainly the colder of the two.

MRS. B.

If you ascertain the temperature of these several bodies by a thermometer (which is a much more accurate test than your feeling), you will find that it is exactly the same.

CAROLINE.

But if they are of the same temperature, why should the one feel colder than the other?

MRS. B.

The hearth and the table feel colder than the carpet or the book, because the latter are not such good *conductors of heat* as the former. Caloric finds a more easy passage through marble and wood, than through leather and worsted; the two former will therefore absorb heat more rapidly from your hand, and consequently give it a stronger sensation of cold than the two latter, although they are all of them really of the same temperature.

CAROLINE.

So, then, the sensation I feel on touching a cold body, is in proportion to the rapidity with which my hand yields its heat to that body?

MRS. B.

Precisely; and, if you lay your hand successively on every object in the room, you will discover which are good, and which are bad conductors of heat, by the different degrees of cold you feel. But, in order to ascertain this point, it is necessary that the several substances should be of the same temperature, which will not be the case with those that are very near the fire, or those that are exposed to a current of cold air from a window or door.

EMILY.

But what is the reason that some bodies are better conductors of heat than others?

MRS. B.

This is a point not well ascertained. It has been conjectured that a certain union or adherence takes place between the caloric and the particles of the body through which it passes. If this adherence be strong, the body detains the heat, and parts with it slowly and reluctantly; if slight, it propagates it freely and rapidly. The conducting power of a body is therefore, inversely, as its tendency to unite with caloric.

EMILY.

That is to say, that the best conductors are those that have the least affinity for caloric.

MRS. B.

Yes; but the term affinity is objectionable in this case, because, as that word is used to express a chemical attraction (which can be destroyed only by decomposition), it cannot be applicable to the slight and transient union that takes place between free caloric and the bodies through which it passes; an union which is so weak, that it constantly yields to the tendency which caloric has to an equilibrium. Now you clearly understand, that the passage of caloric, through bodies that are good conductors, is much more rapid than through those that are bad conductors, and that the former both give and receive it more quickly, and therefore, in a given time, more abundantly, than bad conductors, which makes them feel either hotter or colder, though they may be, in fact, both of the same temperature.

CAROLINE.

Yes, I understand it now; the table, and the book lying upon it, being really of the same temperature, would each receive, in the same space of time, the same quantity of heat from my hand, were their conducting powers equal; but as the table is the best conductor of the two, it will absorb the heat from my hand more rapidly, and consequently produce a stronger sensation of cold than the book.

MRS. B.

Very well, my dear; and observe, likewise, that if you were to heat the table and the book an equal number of degrees above the temperature of your body, the table, which before felt the colder, would now feel the hotter of the two; for, as in the first case it took the heat most rapidly from your hand, so it will now impart heat most rapidly to it. Thus the marble table, which seems to us colder than the mahogany one, will prove the hotter of the two to the ice; for, if it takes heat more rapidly from our hands, which are warmer, it will give out heat more rapidly to the ice, which is colder. Do you understand the reason of these apparently opposite effects?

EMILY.

Perfectly. A body which is a good conductor of caloric, affords it a free passage; so that it penetrates through that body more rapidly than through one which is a bad conductor; and consequently, if it is colder than your hand, you lose more caloric, and if it is hotter, you gain more than with a bad conductor of the same temperature.

MRS. B.

But you must observe that this is the case only when the conductors are either hotter or colder than your hand; for, if you heat different conductors to the temperature of your body, they will all feel equally warm, since the exchange of caloric between bodies of the same temperature is equal. Now, can you tell me why flannel clothing, which is a very bad conductor of heat, prevents our feeling cold?

CAROLINE.

It prevents the cold from penetrating ....

MRS. B.

But you forget that cold is only a negative quality.

CAROLINE.

True; it only prevents the heat of our bodies from escaping so rapidly as it would otherwise do.

MRS. B.

Now you have explained it right; the flannel rather keeps in the heat, than keeps out the cold. Were the atmosphere of a higher temperature than our bodies, it would be equally efficacious in keeping their temperature at the same degree, as it would prevent the free access of the external heat, by the difficulty with which it conducts it.

EMILY.

This, I think, is very clear. Heat, whether external or internal, cannot easily penetrate flannel; therefore in cold weather it keeps us warm; and if the weather was hotter than our bodies, it would keep us cool.

MRS. B.

The most dense bodies are, generally speaking, the best conductors of heat; probably because the denser the body the greater are the number of points or particles that come in contact with caloric. At the common temperature of the atmosphere a piece of metal will feel much colder than a piece of wood, and the latter than a piece of woollen cloth; this again will feel colder than flannel; and down, which is one of the lightest, is at the same time one of the warmest bodies.

CAROLINE.

This is, I suppose, the reason that the plumage of birds preserves them so effectually from the influence of cold in winter?

MRS. B.

Yes; but though feathers in general are an excellent preservative against cold, down is a kind of plumage peculiar to aquatic birds, and covers their chest, which is the part most exposed to the water; for though the surface of the water is not of a lower temperature than the atmosphere, yet, as it is a better conductor of heat, it feels much colder, consequently the chest of the bird requires a warmer covering than any other part of its body. Besides, the breasts of aquatic birds are exposed to cold not only from the temperature of the water, but also from the velocity with which the breast of the bird strikes against it; and likewise from the rapid evaporation occasioned in that part by the air against which it strikes, after it has been moistened by dipping from time to time into the water.

If you hold a finger of one hand motionless in a glass of water, and at the same time move a finger of the other hand swiftly through water of the same temperature, a different sensation will be soon perceived in the different fingers.

Most animal substances, especially those which Providence has assigned as a covering for animals, such as fur, wool, hair, skin, &c. are bad conductors of heat, and are, on that account, such excellent preservatives against the inclemency of winter, that our warmest apparel is made of these materials.

EMILY.

Wood is, I dare say, not so good a conductor as metal, and it is for that reason, no doubt, that silver teapots have always wooden handles.

MRS. B.

Yes; and it is the facility with which metals

conduct caloric that made you suppose that a silver pot radiated more caloric than an earthen one. The silver pot is in fact hotter to the hand when in contact with it; but it is beeause its conducting power more than counterbalances its deficiency in regard to radiation.

We have observed that the most dense bodies are in general the best conductors; and metals, you know, are of that class. Porous bodies, such as the earths and wood, are worse conductors, chiefly, I believe, on account of their pores being filled with air; for air is a remarkably bad conductor.

CAROLINE.

It is a very fortunate circumstance that air should be a bad conductor, as it tends to preserve the heat of the body when exposed to cold weather.

MRS. B.

It is one of the many benevolent dispensations of Providence, in order to soften the inclemency of the seasons, and to render almost all climates habitable to man.

In fluids of different densities, the power of conducting heat varies no less remarkably; if you dip your hand into this vessel full of mercury, you will scarcely conceive that its temperature is not lower than that of the atmosphere.

CAROLINE.

Indeed I know not how to believe it, it feels so extremely cold. — But we may easily ascertain its true temperature by the thermometer. — It is really not colder than the air;—the apparent difference then is produced merely by the difference of the conducting power in mercury and in air.

MRS. B.

Yes; hence you may judge how little the sense of feeling is to be relied on as a test of the temperature of bodies, and how necessary a thermometer is for that purpose.

It has indeed been doubted whether fluids have the power of conducting caloric in the same manner as solid bodies. Count Rumford, a very few years since, attempted to prove, by a variety of experiments, that fluids, when at rest, were not at all endowed with this property.

CAROLINE.

How is that possible, since they are capable of imparting cold or heat to us; for if they did not conduct heat, they would neither take it from, nor give it to us?

MRS. B.

Count Rumford did not mean to say that fluids would not communicate their heat to solid bodies;

but only that heat does not pervade fluids, that is to say, is not transmitted from one particle of a fluid to another, in the same manner as in solid bodies.

EMILY.

But when you heat a vessel of water over the fire, if the particles of water do not communicate heat to each other, how does the water become hot throughout?

MRS. B.

By constant agitation. Water, as you have seen, expands by heat in the same manner as solid bodies; the heated particles of water, therefore, at the bottom of the vessel, become specifically lighter than the rest of the liquid, and consequently ascend to the surface, where, parting with some of their heat to the colder atmosphere, they are condensed, and give way to a fresh succession of heated particles ascending from the bottom, which having thrown off their heat at the surface, are in their turn displaced. Thus every particle is successively heated at the bottom, and cooled at the surface of the liquid; but as the fire communicates heat more rapidly than the atmosphere cools the succession of surfaces, the whole of the liquid in time becomes heated.

CAROLINE.

This accounts most ingeniously for the propagation of heat upwards. But suppose you were to heat the upper surface of a liquid, the particles being specifically lighter than those below, could not descend: how therefore would the heat be communicated downwards?

MRS. B.

If there were no agitation to force the heated surface downwards, Count Rumford assures us that the heat would not descend. In proof of this he succeeded in making the upper surface of a vessel of water boil and evaporate, while a cake of ice remained frozen at the bottom.

CAROLINE.

That is very extraordinary indeed!

MRS. B.

It appears so, because we are not accustomed to heat liquids by their upper surface; but you will understand this theory better if I show you the internal motion that takes place in liquids when they experience a change of temperature. The motion of the liquid itself is indeed invisible from the extreme minuteness of its particles; but if you mix with it any coloured dust, or powder, of nearly the same specific gravity as the liquid, you may judge of the internal motion of the latter by that of the coloured dust it contains.—Do you see the small pieces of amber moving about in the liquid contained in this phial?

CAROLINE.

Yes, perfectly.

MRS. B.

We shall now immerse the phial in a glass of hot water, and the motion of the liquid will be shown, by that which it communicates to the amber.

EMILY.

I see two currents, the one rising along the sides of the phial, the other descending in the centre: but I do not understand the reason of this.

MRS. B.

The hot water communicates its caloric, through the medium of the phial, to the particles of the fluid nearest to the glass; these dilate and ascend laterally to the surface, where, in parting with their heat, they are condensed, and in descending, form the central current.

CAROLINE.

This is indeed a very clear and satisfactory experiment; but how much slower the currents now move than they did at first?

MRS. B.

It is because the circulation of particles has

nearly produced an equilibrium of temperature between the liquid in the glass and that in the phial.

CAROLINE.

But these communicate laterally, and I thought that heat in liquids could be propagated only upwards.

MRS. B.

You do not take notice that the heat is imparted from one liquid to the other, through the medium of the phial itself, the external surface of which receives the heat from the water in the glass, whilst its internal surface transmits it to the liquid it contains. Now take the phial out of the hot water, and observe the effect of its cooling.

EMILY.

The currents are reversed; the external current now descends, and the internal one rises.—I guess the reason of this change:—the phial being in contact with cold air instead of hot water, the external particles are cooled instead of being heated; they therefore descend and force up the central particles, which, being warmer, are consequently lighter.

MRS. B.

It is just so. Count Rumford hence infers that no alteration of temperature can take place in a fluid, without an internal motion of its particles, and as this motion is produced only by the comparative levity of the heated particles, heat cannot be propagated downwards.

But though I believe that Count Rumford's theory as to heat being incapable of pervading fluids is not strictly correct, yet there is, no doubt, much truth in his observation, that the communication is materially promoted by a motion of the parts; and this accounts for the cold that is found to prevail at the bottom of the lakes in Switzerland, which are fed by rivers issuing from the snowy Alps. The water of these rivers being colder, and therefore more dense than that of the lakes, subsides to the bottom, where it cannot be affected by the warmer temperature of the surface; the motion of the waves may communicate this temperature to some little depth, but it can descend no further than the agitation extends.

EMILY.

But when the atmosphere is colder than the lake, the colder surface of the water will descend, for the very reason that the warmer will not.

MRS. B.

Certainly; and it is on this account that neither a lake, nor any body of water whatever, can be frozen until every particle of the water has risen to the surface to give off its caloric to the colder

VOL.I.p.84. PLATE IV

Fig.1.

Pneumatic Pump.

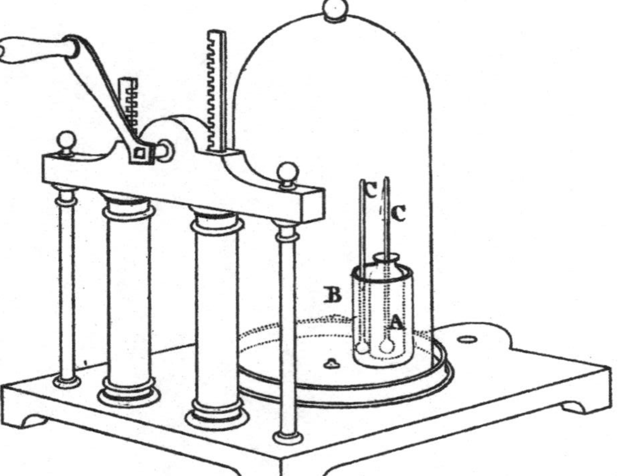

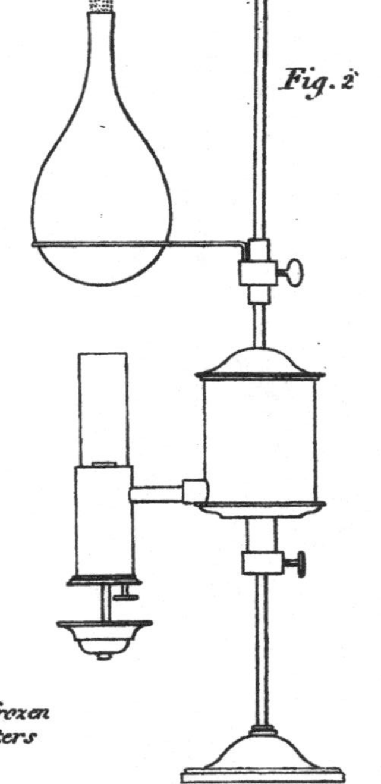

Fig.2.

Fig. 2. Boiling water in a flask over a Patent lamp.—Fig.1. Ether evaporated & water frozen in the air pump.—A Phial of Ether.—B Glass vessel containing water.—C.C Thermometers one in the Ether, the other in the water.

Drawn by the Author. Engraved by Lowry.

Published by Longman & Co. Octr 2nd 1800.

atmosphere; therefore the deeper a body of water is, the longer will be the time it requires to be frozen.

EMILY.

But if the temperature of the whole body of water be brought down to the freezing point, why is only the surface frozen?

MRS. B.

The temperature of the whole body is lowered, but not to the freezing point. The diminution of heat, as you know, produces a contraction in the bulk of fluids, as well as of solids. This effect, however, does not take place in water below the temperature of 40 degrees, which is 8 degrees above the freezing point. At that temperature, therefore, the internal motion, occasioned by the increased specific gravity of the condensed particles, ceases; for when the water at the surface no longer condenses, it will no longer descend, and leave a fresh surface exposed to the atmosphere: this surface alone, therefore, will be further exposed to its severity, and will soon be brought down to the freezing point, when it becomes ice, which being a bad conductor of heat, preserves the water beneath a long time from being affected by the external cold.

CAROLINE.

And the sea does not freeze, I suppose, because

its depth is so great, that a frost never lasts long enough to bring down the temperature of such a great body of water to 40 degrees?

MRS. B.

That is one reason why the sea, as a large mass of water, does not freeze. But, independently of this, salt water does not freeze till it is cooled much below 32 degrees, and with respect to the law of condensation, salt water is an exception, as it condenses even many degrees below the freezing point. When the caloric of fresh water, therefore, is imprisoned by the ice on its surface, the ocean still continues throwing off heat into the atmosphere, which is a most signal dispensation of Providence to moderate the intensity of the cold in winter.

CAROLINE.

This theory of the non-conducting power of liquids, does not, I suppose, hold good with respect to air, otherwise the atmosphere would not be heated by the rays of the sun passing through it?

MRS. B.

Nor is it heated in that way. The pure atmosphere is a perfectly transparent medium, which neither radiates, absorbs, nor conducts caloric, but transmits the rays of the sun to us without in any way diminishing their intensity. The air is therefore not more heated, by the sun's rays passing through it, than diamond, glass, water, or any other transparent medium.

CAROLINE.

That is very extraordinary! Are glass windows not heated then by the sun shining on them?

MRS. B.

No; not if the glass be perfectly transparent. A most convincing proof that glass transmits the rays of the sun without being heated by them is afforded by the burning lens, which by converging the rays to a focus will set combustible bodies on fire, without its own temperature being raised.

EMILY.

Yet, Mrs. B., if I hold a piece of glass near the fire it is almost immediately warmed by it; the glass therefore must retain some of the caloric radiated by the fire? Is it that the solar rays alone pass freely through glass without paying tribute? It seems unaccountable that the radiation of a common fire should have power to do what the sun's rays cannot accomplish.

MRS. B.

It is not because the rays from the fire have more power, but rather because they have less, that they heat glass and other transparent bodies. It is true, however, that as you approach the source of heat the rays being nearer each other, the heat is more condensed, and can produce effects of which the solar rays, from the great distance of their source, are incapable. Thus we should find it impossible to roast a joint of meat by the sun's rays, though it is so easily done by culinary heat. Yet caloric emanated from burning bodies, which is commonly called *culinary heat,* has neither the intensity nor the velocity of solar rays. All caloric, we have said, is supposed to proceed originally from the sun; but after having been incorporated with terrestrial bodies, and again given out by them, though its nature is not essentially altered, it retains neither the intensity nor the velocity with which it first emanated from that luminary; it has therefore not the power of passing through transparent mediums, such as glass and water, without being partially retained by those bodies.

EMILY.

I recollect that in the experiment on the reflection of heat, the glass skreen which you interposed between the burning taper and the mirror, arrested the rays of caloric, and suffered only those of light to pass through it.

CAROLINE.

Glass windows, then, though they cannot be

heated by the sun shining on them, may be heated internally by a fire in the room? But, Mrs. B., since the atmosphere is not warmed by the solar rays passing through it, how does it obtain heat; for all the fires that are burning on the surface of the earth would contribute very little towards warming it?

EMILY.

The radiation of heat is not confined to burning bodies: for all bodies, you know, have that property; therefore, not only every thing upon the surface of the earth, but the earth itself, must radiate heat; and this terrestrial caloric, not having, I suppose, sufficient power to traverse the atmosphere, communicates heat to it.

MRS. B.

Your inference is extremely well drawn, Emily; but the foundation on which it rests is not sound; for the fact is, that terrestrial or culinary heat, though it cannot pass through the denser transparent mediums, such as glass or water, without loss, traverses the atmosphere completely; so that all the heat which the earth radiates, unless it meet with clouds or any foreign body to intercept its passage, passes into the distant regions of the universe.

CAROLINE.

What a pity that so much heat should be wasted!

MRS. B.

Before you are tempted to object to any law of nature, reflect whether it may not prove to be one of the numberless dispensations of Providence for our good. If all the heat which the earth has received from the sun, since the creation had been accumulated in it, its temperature by this time would, no doubt, have been more elevated than any human being could have borne.

CAROLINE.

I spoke indeed very inconsiderately. But, Mrs. B., though the earth, at such a high temperature, might have scorched our feet, we should always have had a cool refreshing air to breathe, since the radiation of the earth does not heat the atmosphere.

EMILY.

The cool air would have afforded but very insufficient refreshment, whilst our bodies were exposed to the burning radiation of the earth.

MRS. B.

Nor should we have breathed a cool air; for though it is true that heat is not communicated to the atmosphere by radiation, yet the air is warmed by contact with heated bodies, in the same manner as solids or liquids. The stratum of air which is immediately in contact with the earth is heated by it; it becomes specifically lighter and rises, making way for another stratum of air which is in its turn heated and carried upwards; and thus each successive stratum of air is warmed by coming in contact with the earth. You may perceive this effect in a sultry day, if you attentively observe the strata of air near the surface of the earth; they appear in constant agitation, for though it is true the air is itself invisible, yet the sun shining on the vapours floating in it, render them visible, like the amber dust in the water. The temperature of the surface of the earth is therefore the source from whence the atmosphere derives its heat, though it is communicated neither by radiation, nor transmitted from one particle of it to another by the conducting power; but every particle of air must come in contact with the earth in order to receive heat from it.

EMILY.

Wind then by agitating the air should contribute to cool the earth and warm the atmosphere, by bringing a more rapid succession of fresh strata of air in contact with the earth, and yet in general wind feels cooler than still air?

MRS. B.

Because the agitation of the air carries off heat from the surface of our bodies more rapidly than still air, by occasioning a greater number of points of contact in a given time.

EMILY.

Since it is from the earth and not the sun that the atmosphere receives its heat, I no longer wonder that elevated regions should be colder than plains and valleys; it was always a subject of astonishment to me, that in ascending a mountain and approaching the sun, the air became colder instead of being more heated.

MRS. B.

At the distance of about a hundred million of miles, which we are from the sun, the approach of a few thousand feet makes no sensible difference, whilst it produces a very considerable effect with regard to the warming the atmosphere at the surface of the earth.

CAROLINE.

Yet as the warm air rises from the earth and the cold air descends to it, I should have supposed that heat would have accumulated in the upper regions of the atmosphere, and that we should have felt the air warmer as we ascended?

MRS. B.

The atmosphere, you know, diminishes in density, and consequently in weight, as it is more distant from the earth; the warm air, therefore, rises only till it meets with a stratum of air of its own density; and it will not ascend into the upper regions of the atmosphere until all the parts beneath have been previously heated. The length of summer even in warm climates does not heat the air sufficiently to melt the snow which has accumulated during the winter on very high mountains, although they are almost constantly exposed to the heat of the sun's rays, being too much elevated to be often enveloped in clouds.

EMILY.

These explanations are very satisfactory; but allow me to ask you one more question respecting the increased levity of heated liquids. You said that when water was heated over the fire, the particles at the bottom of the vessel ascended as soon as heated, in consequence of their specific levity: why does not the same effect continue when the water boils, and is converted into steam? and why does the steam rise from the surface, instead of the bottom of the liquid?

MRS. B.

The steam or vapour does ascend from the bottom, though it seems to arise from the surface of the liquid. We shall boil some water in this Florence flask, (PLATE IV. Fig. 1.) in order that you may be well acquainted with the process of ebullition;—you will then see, through the glass, that the vapour rises in bubbles from the bottom. We shall make it boil by means of a lamp, which is more convenient for this purpose than the chimney fire.

EMILY.

I see some small bubbles ascend, and a great many appear all over the inside of the flask; does the water begin to boil already?

MRS. B.

No; what you now see are bubbles of air, which were either dissolved in the water, or attached to the inner surface of the flask, and which, being rarefied by the heat, ascend in the water.

EMILY.

But the heat which rarefies the air inclosed in the water must rarefy the water at the same time; therefore, if it could remain stationary in the water when both were cold, I do not understand why it should not when both are equally heated?

MRS. B.

Air being much less dense than water, is more easily rarefied; the former, therefore, expands to a great extent, whilst the latter continues to occupy nearly the same space; for water dilates comparatively but very little without changing its state and becoming vapour. Now that the water in the flask begins to boil, observe what large bubbles rise from the bottom of it.

EMILY.

I see them perfectly; but I wonder that they have sufficient power to force themselves through the water.

CAROLINE.

They *must* rise, you know, from their specific levity.

MRS. B.

You are right, Caroline; but vapour has not in all liquids (when brought to the degree of vaporization) the power of overcoming the pressure of the less heated surface. Metals, for instance, mercury excepted, evaporate only from the surface; therefore no vapour will ascend from them till the degree of heat which is necessary to form it has reached the surface; that is to say, till the whole of the liquid is brought to a state of ebullition.

EMILY.

I have observed that steam, immediately issuing from the spout of a teakettle, is less visible than at a further distance from it; yet it must be more dense when it first evaporates, than when it begins to diffuse itself in the air.

MRS. B.

When the steam is first formed, it is so perfectly dissolved by caloric, as to be invisible. In order however to understand this, it will be necessary for me to enter into some explanation respecting the nature of SOLUTION. Solution takes place whenever a body is melted in a fluid. In this operation the body is reduced to such a minute state of division by the fluid, as to become invisible in it, and to partake of its fluidity; but in common solutions this happens without any decomposition, the body being only divided into its integrant particles by the fluid in which it is melted.

CAROLINE.

It is then a mode of destroying the attraction of aggregation.

MRS. B.

Undoubtedly.—The two principal solvent fluids are *water*, and *caloric*. You may have observed that if you melt salt in water, it totally disappears, and the water remains clear, and transparent as before; yet though the union of these two bodies appears so perfect, it is not produced by any chemical combination; both the salt and the water remain unchanged; and if you were to separate them by evaporating the latter, you would find the salt in the same state as before.

EMILY.

I suppose that water is a solvent for solid bodies, and caloric for liquids?

MRS. B.

Liquids of course can only be converted into vapour by caloric. But the solvent power of this agent is not at all confined to that class of bodies; a great variety of solid substances are dissolved by heat: thus metals, which are insoluble in water, can be dissolved by intense heat, being first fused or converted into a liquid, and then rarefied into an invisible vapour. Many other bodies, such as salt, gums, &c. yield to either of these solvents.

CAROLINE.

And that, no doubt, is the reason why hot water will melt them so much better than cold water?

MRS. B.

It is so. Caloric may, indeed, be considered as having, in every instance, some share in the solutoin of a body by water, since water, however low its temperature may be, always contains more or less caloric.

EMILY.

Then, perhaps, water owes its solvent power merely to the caloric contained in it?

MRS. B.

That, probably, would be carrying the speculation too far; I should rather think that water and caloric unite their efforts to dissolve a body, and that the difficulty or facility of effecting this, depend both on the degree of attraction of aggregation to be overcome, and on the arrangement of the particles which are more or less disposed to be divided and penetrated by the solvent.

EMILY.

But have not all liquids the same solvent power as water?

MRS. B.

The solvent power of other liquids varies according to their nature, and that of the substances submitted to their action. Most of these solvents, indeed, differ essentially from water, as they do not merely separate the integrant particles of the bodies which they dissolve, but attack their constituent principles by the power of chemical attraction, thus producing a true decomposition. These more complicated operations we must consider in another place, and confine our attention at present to the solutions by water and caloric.

CAROLINE.

But there are a variety of substances which, when dissolved in water, make it thick and muddy, and destroy its transparency.

MRS. B.

In this case it is not a solution, but simply a mixture. I shall show you the difference between a solution and a mixture, by putting some common salt into one glass of water, and some powder of chalk into another; both these substances are white, but their effect on the water will be very different.

CAROLINE.

Very different indeed! The salt entirely disappears and leaves the water transparent, whilst the chalk changes it into an opaque liquid like milk.

EMILY.

And would lumps of chalk and salt produce similar effects on water?

MRS. B.

Yes, but not so rapidly; salt is, indeed, soon melted though in a lump; but chalk, which does not mix so readily with water, would require a much greater length of time; I therefore preferred showing you the experiment with both substances reduced to powder, which does not in any respect alter their nature, but facilitates the operation merely by presenting a greater quantity of surface to the water.

I must not forget to mention a very curious circumstance respecting solutions, which is, that a fluid is not nearly so much increased in bulk by holding a body in solution, as it would by mere mixture with the body.

CAROLINE.

That seems impossible; for two bodies cannot exist together in the same space.

MRS. B.

Two bodies may, by condensation, occupy less space when in union than when separate, and this I can show you by an easy experiment.

This phial, which contains some salt, I shall fill with water, pouring it in quickly, so as not to dissolve much of the salt; and when it is quite full I cork it.—If I now shake the phial till the salt is dissolved, you will observe that it is no longer full.

CAROLINE.

I shall try to add a little more salt.—But now, you see, Mrs. B., the water runs over.

MRS. B.

Yes; but observe that the last quantity of salt you put in remains solid at the bottom, and displaces the water; for it has already melted all the salt it is capable of holding in solution. This is called the point of *saturation*; and the water in this case is said to be *saturated* with salt.

EMILY.

I think I now understand the solution of a solid body by water perfectly: but I have not so clear an idea of the solution of a liquid by caloric.

MRS. B.

It is probably of a similar nature; but as caloric is an invisible fluid, its action as a solvent is not so obvious as that of water. Caloric, we may conceive, dissolves water, and converts it into vapour by the same process as water dissolves salt; that is to say, the particles of water are so minutely divided by the caloric as to become invisible. Thus, you are now enabled to understand why the vapour of boiling water, when it first issues from the spout of a kettle, is invisible; it is so, because it is then completely dissolved by caloric. But the air with which it comes in contact, being much colder than the vapour, the latter yields to it a quantity of its caloric. The particles of vapour being thus in a great measure deprived of their solvent, gradually collect, and become visible in the form of steam, which is water in a state of imperfect solution; and if you were further to deprive it of its caloric, it would return to its original liquid state.

CAROLINE.

That I understand very well. If you hold a cold plate over a tea-urn, the steam issuing from it will be immediately converted into drops of water by parting with its caloric to the plate; but in what state is the steam, when it becomes invisible by being diffused in the air?

MRS. B.

It is not merely diffused, but is again dissolved by the air.

EMILY.

The air, then, has a solvent power, like water and caloric?

MRS. B.

This was formerly believed to be the case. But it appears from more recent enquiries that the solvent power of the atmosphere depends solely upon the caloric contained in it. Sometimes the watery vapour diffused in the atmosphere is but imperfectly dissolved, as is the case in the formation of clouds and fogs; but if it gets into a region sufficiently warm, it becomes perfectly invisible.

EMILY.

Can any water dissolve in the atmosphere without its being previously converted into vapour by boiling?

MRS. B.

Unquestionably; and this constitutes the difference between *vaporization* and *evaporation.* Water, when heated to the boiling point, can no longer exist in the form of water, and must necessarily be converted into vapour or steam, whatever may be the state and temperature of the surrounding medium; this is called vaporization. But the atmosphere, by means of the caloric it contains, can take up a certain portion of water at any temperature, and hold it in a state of solution. This is simply evaporation. Thus the atmosphere is continually carrying off moisture from the surface of the earth, until it is saturated with it.

CAROLINE.

That is the case, no doubt, when we feel the atmosphere damp.

MRS. B.

On the contrary, when the moisture is well dissolved it occasions no humidity: it is only when in a state of imperfect solution and floating in the atmosphere, in the form of watery vapour, that it produces dampness. This happens more frequently in winter than in summer; for the lower the temperature of the atmosphere, the less water it can dissolve; and in reality it never contains so much moisture as in a dry hot summer's day.

CAROLINE.

You astonish me! But why, then, is the air so dry in frosty weather, when its temperature is at the lowest?

EMILY.

This, I conjecture, proceeds not so much from the moisture being dissolved, as from its being frozen; is not that the case?

MRS. B.

It is; and the freezing of the watery vapour which the atmospheric heat could not dissolve, produces what is called a hoar frost; for the particles descend in freezing, and attach themselves to whatever they meet with on the surface of the earth.

The tendency of free caloric to an equilibrium, together with its solvent power, are likewise connected with the phenomena of rain, of dew, &c. When moist air of a certain temperature happens to pass through a colder region of the atmosphere, it parts with a portion of its heat to the surrounding air; the quantity of caloric, therefore, which served to keep the water in a state of vapour, being diminished, the watery particles approach each other, and form themselves into drops of water, which being heavier than the atmosphere, descend to the earth. There are also other circumstances, and particularly the variation in the weight of the atmosphere, which may contribute to the formation of rain. This, however, is an intricate subject, into which we cannot more fully enter at present.

EMILY.

In what manner do you account for the formation of dew?

MRS. B.

Dew is a deposition of watery particles or minute drops from the atmosphere, precipitated by the coolness of the evening.

CAROLINE.

This precipitation is owing, I suppose, to the cooling of the atmosphere, which prevents its retaining so great a quantity of watery vapour in solution as during the heat of the day.

MRS. B.

Such was, from time immemorial, the generally received opinion respecting the cause of dew; but it has been very recently proved by a course of ingenious experiments of Dr. Wells, that the deposition of dew is produced by the cooling of the surface of the earth, which he has shown to take place previously to the cooling of the atmosphere; for on examining the temperature of a plot of grass just before the dew-fall, he found that is was considerably colder than the air a few feet above it, from which the dew was shortly after precipitated.

EMILY.

But why should the earth cool in the evening sooner than the atmosphere?

MRS. B.

Because it parts with its heat more readily than the air; the earth is an excellent radiator of caloric, whilst the atmosphere does not possess that property, at least in any sensible degree. Towards evening, therefore, when the solar heat declines, and when after sunset it entirely ceases, the earth rapidly cools by radiating heat towards the skies; whilst the air has no means of parting with its heat but by coming into contact with the cooled surface of the earth, to which it communicates its caloric. Its solvent power being thus reduced, it is unable to retain so large a portion of watery vapour, and deposits those pearly drops which we call dew.

EMILY.

If this be the cause of dew, we need not be apprehensive of receiving any injury from it; for it can be deposited only on surfaces that are colder than the atmosphere, which is never the case with our bodies.

MRS. B.

Very true; yet I would not advise you for this reason to be too confident of escaping all the ill effects which may arise from exposure to the dew; for it may be deposited on your clothes, and chill you afterwards by its evaporation from them. Besides, whenever the dew is copious, there is a chill in the atmosphere which it is not always safe to encounter.

CAROLINE.

Wind, then, must promote the deposition of dew, by bringing a more rapid succession of particles of air in contact with the earth, just as it promotes the cooling of the earth and warming of the atmosphere during the heat of the day?

MRS. B.

Yes; provided the wind be unattended with clouds, for these accumulations of moisture not only prevent the free radiation of the earth towards the upper regions, but themselves radiate towards the earth; under these circumstances much less dew is formed than on fine clear nights, when the radiation of the earth passes without obstacle through the atmosphere to the distant regions of space, whence it receives no caloric in exchange. The dew continues to be deposited during the night; and is generally most abundant towards morning, when the contrast between the temperature of the earth and that of the air is greatest. After sunrise the equilibrium of temperature between these two bodies is gradually restored by the solar rays passing freely through the atmosphere to the earth; and later in the morning the temperature of the earth gains the ascendency, and gives out caloric to the air by contact, in the same manner as it receives it from the air during the night.— Can you tell me, now, why a bottle of wine taken fresh from the cellar (in summer particularly), will soon be covered with dew; and even the glasses into which the wine is poured will be moistened with a similar vapour?

EMILY.

The bottle being colder than the surrounding air must absorb caloric from it; the moisture therefore which that air contained becomes visible, and forms the dew which is deposited on the bottle.

MRS. B.

Very well, Emily. Now, Caroline, can you inform me why, in a warm room, or close carriage, the contrary effect takes place; that is to say, that the inside of the windows is covered with vapour?

CAROLINE.

I have heard that it proceeds from the breath of those within the room or the carriage; and I suppose it is occasioned by the windows which, being colder than the breath, deprive it of part of its caloric, and by this means convert it into watery vapour.

MRS. B.

You have both explained it extremely well. Bodies attract dew in proportion as they are good radiators of caloric, as it is this quality which reduces their temperature below that of the atmosphere; hence we find that little or no dew is deposited on rocks, sand, water; while grass and living vegetables, to which it is so highly beneficial, attract it in abundance—another remarkable instance of the wise and bountiful dispensations of Providence.

EMILY.

And we may again observe it in the abundance of dew in summer, and in hot climates, when its cooling effects are so much required; but I do not understand what natural cause increases the dew in hot weather?

MRS. B.

The more caloric the earth receives during the day, the more it will radiate afterwards, and consequently the more rapidly its temperature will be reduced in the evening, in comparison to that of the atmosphere. In the West-Indies especially, where the intense heat of the day is strongly contrasted with the coolness of the evening, the dew is prodigiously abundant. During a drought, the dew is less plentiful, as the earth is not sufficiently supplied with moisture to be able to saturate the atmosphere.

CAROLINE.

I have often observed, Mrs. B., that when I walk out in frosty weather, with a veil over my face, my breath freezes upon it. Pray what is the reason of that?

MRS. B.

It is because the cold air immediately seizes on the caloric of your breath, and, by robbing it of its solvent, reduces it to a denser fluid, which is the watery vapour that settles on your veil, and there it continues parting with its caloric till it is brought down to the temperature of the atmosphere, and assumes the form of ice.

You may, perhaps, have observed that the breath of animals, or rather the moisture contained in it, is visible in damp weather, or during a frost. In the former case, the atmosphere being oversaturated with moisture, can dissolve no more. In the latter, the cold condenses it into visible vapour; and for the same reason, the steam arising from water that is warmer than the atmosphere, becomes visible. Have you never taken notice of the vapour rising from your hands after having dipped them into warm water?

CAROLINE.

Frequently, especially in frosty weather.

MRS. B.

We have already observed that pressure is an obstacle to evaporation: there are liquids that contain so great a quantity of caloric, and whose particles consequently adhere so slightly together, that they may be rapidly converted into vapour without any elevation of temperature, merely by taking off the weight of the atmosphere. In such liquids, you perceive, it is the pressure of the atmosphere alone that connects their particles, and keeps them in a liquid state.

CAROLINE.

I do not well understand why the particles of such fluids should be disunited and converted into vapour, without any elevation of temperature, in spite of the attraction of cohesion.

MRS. B.

It is because the degree of heat at which we usually observe these fluids is sufficient to overcome their attraction of cohesion. Ether is of this description; it will boil and be converted into vapour, at the common temperature of the air, if the pressure of the atmosphere be taken off.

EMILY.

I thought that ether would evaporate without either the pressure of the atmosphere being taken away, or heat applied; and that it was for that reason so necessary to keep it carefully corked up?

MRS. B.

It is true it will evaporate, but without ebullition; what I am now speaking of is the vaporization of ether, or its conversion into vapour by boiling. I am going to show you how suddenly the ether in this phial will be converted into vapour, by means of the air-pump. — Observe with what rapidity the bubbles ascend, as I take off the pressure of the atmosphere.

CAROLINE.

It positively boils: how singular to see a liquid boil without heat!

MRS. B.

Now I shall place the phial of ether in this glass, which it nearly fits, so as to leave only a small space, which I fill with water; and in this state I put it again under the receiver. (PLATE

IV. Fig. 1.)*—You will observe, as I exhaust the air from it, that whilst the ether boils, the water freezes.

CAROLINE.

It is indeed wonderful to see water freeze in contact with a boiling fluid!

EMILY.

I am at a loss to conceive how the ether can pass to the state of vapour without an addition of caloric. Does it not contain more caloric in a state of vapour, than in a state of liquidity?

MRS. B.

It certainly does; for though it is the pressure of the atmosphere which condenses it into a liquid, it is by forcing out the caloric that belongs to it when in an aëriform state.

* Two pieces of thin glass tubes, sealed at one end, might answer this purpose better. The experiment, however, as here described, is difficult, and requires a very nice apparatus. But if, instead of phials or tubes, two watch-glasses be used, water may be frozen almost instantly in the same manner. The two glasses are placed over one another, with a few drops of water interposed between them, and the uppermost glass is filled with ether. After working the pump for a minute or two, the glasses are found to adhere strongly together, and a thin layer of ice is seen between them.

EMILY.

You have, therefore, two difficulties to explain, Mrs. B.—First, from whence the ether obtains the caloric necessary to convert it into vapour when it is relieved from the pressure of the atmosphere; and, secondly, what is the reason that the water, in which the bottle of ether stands, is frozen?

CAROLINE.

Now, I think, I can answer both these questions. The ether obtains the addition of caloric required, from the water in the glass; and the loss of caloric, which the latter sustains, is the occasion of its freezing.

MRS. B.

You are perfectly right; and if you look at the thermometer which I have placed in the water, whilst I am working the pump, you will see that every time bubbles of vapour are produced, the mercury descends; which proves that the heat of the water diminishes in proportion as the ether boils.

EMILY.

This I understand now very well; but if the water freezes in consequence of yielding its caloric to the ether, the equilibrium of heat must, in this case, be totally destroyed. Yet you have told us, that the exchange of caloric between two bodies of equal temperature, was always equal; how, then, is it that the water, which was originally of the same temperature as the ether, gives out caloric to it, till the water is frozen, and the ether made to boil?

MRS. B.

I suspected that you would make these objections; and, in order to remove them, I enclosed two thermometers in the air-pump; one which stands in the glass of water, the other in the phial of ether; and you may see that the equilibrium of temperature is not destroyed; for as the thermometer descends in the water, that in the ether sinks in the same manner; so that both thermometers indicate the same temperature, though one of them is in a boiling, the other in a freezing liquid.

EMILY.

The ether, then, becomes colder as it boils? This is so contrary to common experience, that I confess it astonishes me exceedingly.

CAROLINE.

It is, indeed, a most extraordinary circumstance. But pray, how do you account for it?

MRS. B.

I cannot satisfy your curiosity at present; for before we can attempt to explain this apparent paradox, it is necessary to become acquainted with the subject of LATENT HEAT: and that, I think, we must defer till our next interview.

CAROLINE.

I believe, Mrs. B., that you are glad to put off the explanation; for it must be a very difficult point to account for.

MRS. B.

I hope, however, that I shall do it to your complete satisfaction.

EMILY.

But before we part, give me leave to ask you one question. Would not water, as well as ether, boil with less heat, if deprived of the pressure of the atmosphere?

MRS. B.

Undoubtedly. You must always recollect that there are two forces to overcome, in order to make a liquid boil or evaporate; the attraction of aggregation, and the weight of the atmosphere. On the summit of a high mountain (as Mr. De Saussure ascertained on Mount Blanc) much less heat is required to make water boil, than in the plain, where the weight of the atmosphere is greater. * Indeed if the weight of the atmosphere be entirely removed by means of a good air-pump, and if water be placed in the exhausted receiver, it will evaporate so fast, however cold it may be, as to give it the appearance of boiling from the surface. But without the assistance of the air-pump, I can show you a very pretty experiment, which proves the effect of the pressure of the atmosphere in this respect.

Observe, that this Florence flask is about half full of water, and the upper half of invisible vapour, the water being in the act of boiling. — I take it from the lamp, and cork it carefully—the water, you see, immediately ceases boiling. — I shall now dip the flask into a bason of cold water. †

CAROLINE.

But look, Mrs. B., the hot water begins to boil again, although the cold water must rob it more and more of its caloric! What can be the reason of that?

* On the top of Mount Blanc, water boiled when heated only to 187 degrees, instead of 212 degrees.

† The same effect may be produced by wrapping a cold wet linen cloth round the upper part of the flask. In order to show how much the water cools whilst it is boiling, a thermometer, graduated on the tube itself, may be introduced into the bottle through the cork.

MRS. B.

Let us examine its temperature. You see the thermometer immersed in it remains stationary at 180 degrees, which is about 30 degrees below the boiling point. When I took the flask from the lamp, I observed to you that the upper part of it was filled with vapour; this being compelled to yield its caloric to the cold water, was again condensed into water—What, then, filled the upper part of the flask?

EMILY.

Nothing; for it was too well corked for the air to gain admittance, and therefore the upper part of the flask must be a vacuum.

MRS. B.

The water below, therefore, no longer sustains the pressure of the atmosphere, and will consequently boil at a much lower temperature. Thus, you see, though it had lost many degrees of heat, it began boiling again the instant the vacuum was formed above it. The boiling has now ceased, the temperature of the water being still farther reduced; if it had been ether, instead of water, it would have continued boiling much longer, for ether boils, under the usual atmospheric pressure, at a temperature as low as 100 degrees; and in a vacuum it boils at almost any temperature; but water being a more dense fluid, requires a more considerable quantity of caloric to make it evaporate quickly, even when the pressure of the atmosphere is removed.

EMILY.

What proportion of vapour can the atmosphere contain in a state of solution?

MRS. B.

I do not know whether it has been exactly ascertained by experiment; but at any rate this proportion must vary, both according to the temperature and the weight of the atmosphere; for the lower the temperature, and the greater the pressure, the smaller must be the proportion of vapour that the atmosphere can contain.

To conclude the subject of free caloric, I should mention *Ignition*, by which is meant that emission of light which is produced in bodies at a very high temperature, and which is the effect of accumulated caloric.

EMILY.

You mean, I suppose, that light which is produced by a burning body?

MRS. B.

No: ignition is quite independent of combustion. Clay, chalk, and indeed all incombustible substances, may be made red hot. When a body burns, the light emitted is the effect of a chemical change which takes place, whilst ignition is the effect of caloric alone, and no other change than that of temperature is produced in the ignited body.

All solid bodies, and most liquids, are susceptible of ignition, or, in other words, of being heated so as to become luminous; and it is remarkable that this takes place pretty nearly at the same temperature in all bodies, that is, at about 800 degrees of Fahrenheit's scale.

EMILY.

But how can liquids attain so high a temperature, without being converted into vapour?

MRS. B.

By means of confinement and pressure. Water confined in a strong iron vessel (called Papin's digester) can have its temperature raised to upwards of 400 degrees. Sir James Hall has made some very curious experiments on the effects of heat assisted by pressure; by means of strong gun-barrels, he succeeded in melting a variety of substances which were considered as infusible: and it is not unlikely that, by similar methods, water itself might be heated to redness.

EMILY.

I am surprised at that: for I thought that the force of steam was such as to destroy almost all mechanical resistance.

MRS. B.

The expansive force of steam is prodigious; but in order to subject water to such high temperatures, it is prevented by confinement from being converted into steam, and the expansion of heated water is comparatively trifling.—But we have dwelt so long on the subject of free caloric, that we must reserve the other modifications of that agent to our next meeting, when we shall endeavour to proceed more rapidly.

CONVERSATION IV.

ON COMBINED CALORIC, COMPREHENDING SPECIFIC AND LATENT HEAT.

MRS. B.

WE are now to examine the other modifications of caloric.

CAROLINE.

I am very curious to know of what nature they can be; for I have no notion of any kind of heat that is not perceptible to the senses.

MRS. B.

In order to enable you to understand them, it will be necessary to enter into some previous explanations.

It has been discovered by modern chemists, that bodies of a different nature, heated to the same temperature, do not contain the same quantity of caloric.

CAROLINE.

How could that be ascertained? Have you not told us that it is impossible to discover the absolute quantity of caloric which bodies contain?

MRS. B.

True; but at the same time I said that we were enabled to form a judgment of the proportions which bodies bore to each other in this respect. Thus it is found that, in order to raise the temperature of different bodies the same number of degrees, different quantities of caloric are required for each of them. If, for instance, you place a pound of lead, a pound of chalk, and a pound of milk, in a hot oven, they will be gradually heated to the temperature of the oven; but the lead will attain it first, the chalk next, and the milk last.

CAROLINE.

That is a natural consequence of their different bulks; the lead being the smallest body, will be heated soonest, and the milk, which is the largest, will require the longest time.

MRS. B.

That explanation will not do, for if the lead be the least in bulk, it offers also the least surface to the caloric, the quantity of heat therefore which can enter into it in the same space of time is proportionally smaller.

EMILY.

Why, then, do not the three bodies attain the temperature of the oven at the same time?

MRS. B.

It is supposed to be on account of the different capacity of these bodies for caloric.

CAROLINE.

What do you mean by the *capacity* of a body for caloric?

MRS. B.

I mean a certain disposition of bodies to require more or less caloric for raising their temperature to any degree of heat. Perhaps the fact may be thus explained:

Let us put as many marbles into this glass as it will contain, and pour some sand over them — observe how the sand penetrates and lodges between them. We shall now fill another glass with pebbles of various forms — you see that they arrange themselves in a more compact manner than the marbles, which, being globular, can touch each other by a single point only. The pebbles, therefore, will not admit so much sand between them; and consequently one of these glasses will necessarily contain more sand than the other, though both of them be equally full.

CAROLINE.

This I understand perfectly. The marbles and the pebbles represent two bodies of different kinds, and the sand the caloric contained in them; and it appears very plain, from this comparison, that one body may admit of more caloric between its particles than another.

MRS. B.

You can no longer be surprised, therefore, that bodies of a different capacity for caloric should require different proportions of that fluid to raise their temperatures equally.

EMILY.

But I do not conceive why the body that contains the most caloric should not be of the highest temperature; that is to say, feel hot in proportion to the quantity of caloric it contains?

MRS. B.

The caloric that is employed in filling the capa city of a body, is not free caloric; but is imprisoned as it were in the body, and is therefore imperceptible: for we can feel only the caloric which the body parts with, and not that which it retains.

CAROLINE.

It appears to me very extraordinary that heat should be confined in a body in such a manner as to be imperceptible.

MRS. B.

If you lay your hand on a hot body, you feel only the caloric which leaves it, and enters your hand; for it is impossible that you should be sensible of that which remains in the body. The thermometer, in the same manner, is affected only by the free caloric which a body transmits to it, and not at all by that which it does not part with.

CAROLINE.

I begin to understand it: but I confess that the idea of insensible heat is so new and strange to me, that it requires some time to render it familiar.

MRS. B.

Call it insensible caloric, and the difficulty will appear much less formidable. It is indeed a sort of contradiction to call it heat, when it is so situated as to be incapable of producing that sensation. Yet this modification of caloric is commonly called SPECIFIC HEAT.

CAROLINE.

But it certainly would have been more correct to have called it *specific caloric.*

EMILY.

I do not understand how the term *specific* applies to this modification of caloric?

MRS. B.

It expresses the relative quantity of caloric which different *species* of bodies of the same weight and temperature are capable of containing. This modification is also frequently called *heat of capacity*, a term perhaps preferable, as it explains better its own meaning.

You now understand, I suppose, why the milk and chalk required a longer portion of time than the lead to raise their temperature to that of the oven?

EMILY.

Yes: the milk and chalk having a greater capacity for caloric than the lead, a greater proportion of that fluid became insensible in those bodies: and the more slowly, therefore, their temperature was raised.

CAROLINE.

But might not this difference proceed from the different conducting powers of heat in these three bodies, since that which is the best conductor must necessarily attain the temperature of the oven first?

MRS. B.

Very well observed, Caroline. This objection would be insurmountable, if we could not, by reversing the experiment, prove that the milk, the chalk, and the lead, actually absorbed different quantities of caloric, and we know that if the different time they took in heating, proceeded merely from their different conducting powers, they would each have acquired an equal quantity of caloric.

CAROLINE.

Certainly. But how can you reverse this experiment?

MRS. B.

It may be done by cooling the several bodies to the same degree in an apparatus adapted to receive and measure the caloric which they give out. Thus, if you plunge them into three equal quantities of water, each at the same temperature, you will be able to judge of the relative quantity of caloric which the three bodies contained, by that, which, in cooling, they communicated to their respective portions of water: for the same quantity of caloric which they each absorbed to raise their temperature, will abandon them in lowering it; and on examining the three vessels of water, you will find the one in which you immersed the lead to be the least heated; that which held the chalk will be the next; and that which contained the milk will be heated the most of all. The celebrated Lavoisier has invented a machine to estimate, upon this principle, the specific heat of bodies in a more perfect manner; but I cannot explain it to you, till you are acquainted with the next modification of caloric.

EMILY.

The more dense a body is, I suppose, the less is its capacity for caloric?

MRS. B.

This is not always the case with bodies of different nature; iron, for instance, contains more specific heat than tin, though it is more dense. This seems to show that specific heat does not merely depend upon the interstices between the particles; but, probably, also upon some peculiar constitution of the bodies which we do not comprehend.

EMILY.

But, Mrs. B., it would appear to me more proper to compare bodies by *measure*, rather than by *weight*, in order to estimate their specific heat. Why, for instance, should we not compare *pints* of milk, of chalk, and of lead, rather than *pounds* of those substances; for equal weights may be composed of very different quantities?

MRS. B.

You are mistaken, my dear; equal weight must contain equal quantities of matter; and when we wish to know what is the relative quantity of caloric, which substances of various kinds are capable of containing under the same temperature, we must compare equal weights, and not equal bulks of those substances. Bodies of the same weight may undoubtedly be of very different dimensions; but that does not change their real quantity of matter. A pound of feathers does not contain one atom more than a pound of lead.

CAROLINE.

I have another difficulty to propose. It appears to me, that if the temperature of the three bodies in the oven did not rise equally, they would never reach the same degree; the lead would always keep its advantage over the chalk and milk, and would perhaps be boiling before the others had attained the temperature of the oven. I think you might as well say that, in the course of time, you and I should be of the same age?

MRS. B.

Your comparison is not correct, Caroline. As soon as the lead reached the temperature of the oven, it would remain stationary; for it would then give out as much heat as it would receive. You should recollect that the exchange of radiating heat, between two bodies of equal temperature, is equal: it would be impossible, therefore, for the lead to accumulate heat after having attained the temperature of the oven; and that of the chalk and milk therefore would ultimately arrive at the same standard. Now I fear that this will not hold good with respect to our ages, and that, as long as I live, I shall never cease to keep my advantage over you.

EMILY.

I think that I have found a comparison for specific heat, which is very applicable. Suppose that two men of equal weight and bulk, but who required different quantities of food to satisfy their appetites, sit down to dinner, both equally hungry; the one would consume a much greater quantity of provisions than the other, in order to be equally satisfied.

MRS. B.

Yes, that is very fair; for the quantity of food necessary to satisfy their respective appetites, varies in the same manner as the quantity of caloric requisite to raise equally the temperature of different bodies.

EMILY.

The thermometer, then, affords no indication of the specific heat of bodies?

MRS. B.

None at all: no more than satiety is a test of the quantity of food eaten. The thermometer, as

I have repeatedly said, can be affected only by free caloric, which alone raises the temperature of bodies.

But there is another mode of proving the existence of specific heat, which affords a very satisfactory illustration of that modification. This, however, I did not enlarge upon before, as I thought it might appear to you rather complicated. — If you mix two fluids of different temperatures, let us say the one at 50 degrees, and the other at 100 degrees, of what temperature do you suppose the mixture will be?

CAROLINE.

It will be no doubt the medium between the two, that is to say, 75 degrees.

MRS. B.

That will be the case if the two bodies happen to have the same capacity for caloric; but if not, a different result will be obtained. Thus, for instance, if you mix together a pound of mercury, heated at 50 degrees, and a pound of water heated at 100 degrees, the temperature of the mixture, instead of being 75 degrees, will be 80 degrees; so that the water will have lost only 12 degrees, whilst the mercury will have gained 38 degrees; from which you will conclude that the capacity of mercury for heat is less than that of water.

CAROLINE.

I wonder that mercury should have so little specific heat. Did we not see it was a much better conductor of heat than water?

MRS. B.

And it is precisely on that account that its specific heat is less. For since the conductive power of bodies depends, as we have observed before, on their readiness to receive heat and part with it, it is natural to expect that those bodies which are the worst conductors should absorb the most caloric before they are disposed to part with it to other bodies. But let us now proceed to LATENT HEAT.

CAROLINE.

And pray what kind of heat is that?

MRS. B.

It is another modification of combined caloric, which is so analogous to specific heat, that most chemists make no distinction between them; but Mr. Pictet, in his Essay on Fire, has so clearly discriminated them, that I am induced to adopt his view of the subject. We therefore call *latent heat* that portion of insensible caloric which is employed in changing the state of bodies; that is to say, in converting solids into liquids, or liquids into vapour. When a body changes its state from solid to liquid, or from liquid to vapour, its expansion occasions a sudden and considerable increase of capacity for heat, in consequence of which it immediately absorbs a quantity of caloric, which becomes fixed in the body which it has transformed; and, as it is perfectly concealed from our senses, it has obtained the name of *latent* heat.

CAROLINE.

I think it would be much more correct to call this modification latent caloric instead of latent heat, since it does not excite the sensation of heat.

MRS. B.

This modification of heat was discovered and named by Dr. Black long before the French chemists introduced the term caloric, and we must not presume to alter it, as it is still used by much better chemists than ourselves. And, besides, you are not to suppose that the nature of heat is altered by being variously modified: for if latent heat and specific heat do not excite the same sensations as free caloric, it is owing to their being in a state of confinement, which prevents them from acting upon our organs; and consequently, as soon as they are extricated from the body in which they are imprisoned, they return to their state of free caloric.

EMILY.

But I do not yet clearly see in what respect latent heat differs from specific heat; for they are both of them imprisoned and concealed in bodies.

MRS. B.

Specific heat is that which is employed in filling the capacity of a body for caloric, in the state in which this body actually exists; while latent heat is that which is employed only in effecting a change of state, that is, in converting bodies from a solid to a liquid, or from a liquid to an aëriform state. But I think that, in a general point of view, both these modifications might be comprehended under the name of *heat of capacity*, as in both cases the caloric is equally engaged in filling the capacities of bodies.

I shall now show you an experiment, which I hope will give you a clear idea of what is understood by latent heat.

The snow which you see in this phial has been cooled by certain chemical means (which I cannot well explain to you at present), to 5 or 6 degrees below the freezing point, as you will find indicated by the thermometer which is placed in it. We shall expose it to the heat of a lamp, and you will see the thermometer gradually rise, till it reaches the freezing point ——

EMILY.

But there it stops, Mrs. B., and yet the lamp burns just as well as before. Why is not its heat communicated to the thermometer?

CAROLINE.

And the snow begins to melt, therefore it must be rising above the freezing point?

MRS. B.

The heat no longer affects the thermometer, because it is wholly employed in converting the ice into water. As the ice melts, the caloric becomes *latent* in the new-formed liquid, and therefore cannot raise its temperature; and the thermometer will consequently remain stationary, till the whole of the ice be melted.

CAROLINE.

Now it is all melted, and the thermometer begins to rise again.

MRS. B.

Because the conversion of the ice into water being completed, the caloric no longer becomes latent; and therefore the heat which the water now receives raises its temperature, as you find the thermometer indicates.

EMILY.

But I do not think that the thermometer rises so quickly in the water as it did in the ice, previous to its beginning to melt, though the lamp burns equally well?

MRS. B.

That is owing to the different specific heat of ice and water. The capacity of water for caloric being greater than that of ice, more heat is required to raise its temperature, and therefore the thermometer rises slower in the water than in the ice.

EMILY.

True; you said that a solid body always increased its capacity for heat by becoming fluid; and this is an instance of it.

MRS. B.

Yes, and the latent heat is that which is absorbed in consequence of the greater capacity which the water has for heat, in comparison to ice.

I must now tell you a curious calculation founded on that consideration. I have before observed to you that though the thermometer shows us the comparative warmth of bodies, and enables us to determine the same point at different times and places, it gives us no idea of the absolute quantity of heat in any body. We cannot tell how low it ought to fall by the privation of all heat, but an attempt has been made to infer it in the following manner. It has been found by experiment, that the capacity of water for heat, when compared with that of ice, is as 10 to 9, so that, at the same temperature, ice contains one tenth of caloric less than water. By experiment also it is observed, that in order to melt ice, there must be added to it as much heat, as would, if it did not melt it, raise its temperature 140 degrees. This quantity of heat is therefore absorbed when the ice, by being converted into water, is made to contain one-ninth more caloric than it did before. Therefore 140 degrees is a ninth part of the heat contained in ice at 32 degrees; and the point of zero, or the absolute privation of heat, must consequently be 1260 degrees below 32 degrees.

This mode of investigating so curious a question is ingenious, but its correctness is not yet established by similar calculations for other bodies. The points of absolute cold, indicated by this method in various bodies, are very remote from each other; it is however possible, that this may arise from some imperfection in the experiments.

CAROLINE.

It is indeed very ingenious — but we must now attend to our present experiment. The water begins to boil, and the thermometer is again stationary.

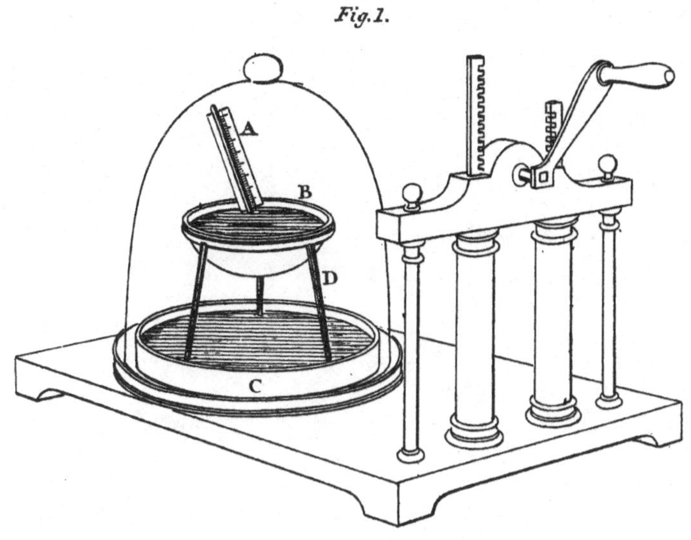

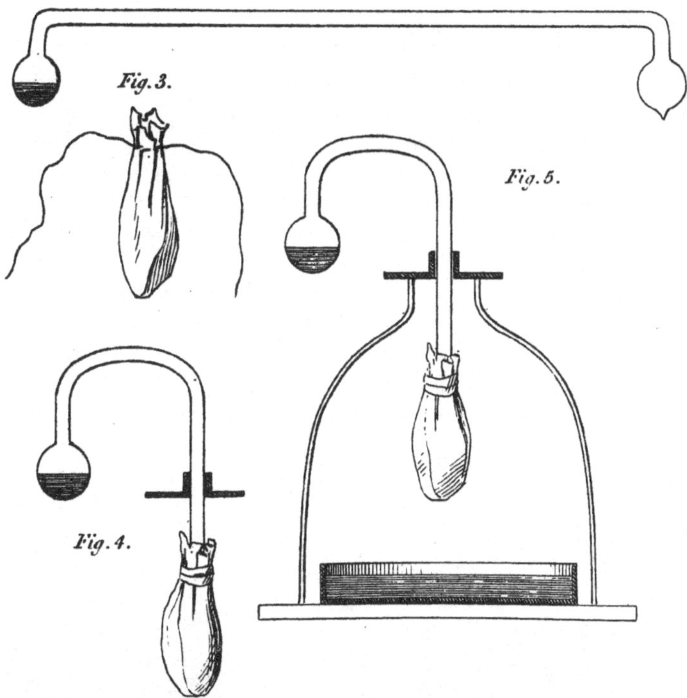

Fig. 1. The air-pump & receiver for Mr. Leslie's experiment, C a saucer with sulphuric Acid, B a glass or earthen cup containing Water. D a stand for the cup with its legs made of Glass. A a Thermometer.
Fig. 2. Dr. Wollaston's Cryophorus, Fig. 5. Dr. Marcet's mode of using the Cryophorus.
Fig. 3. & 4. the different parts of Fig. 5. seen separate.

Lowry sculp.

MRS. B.

Well, Caroline, it is your turn to explain the phenomenon.

CAROLINE.

It is wonderfully curious! The caloric is now busy in changing the water into steam, in which it hides itself, and becomes insensible. This is another example of latent heat, producing a change of form. At first it converted a solid body into a liquid, and now it turns the liquid into vapour!

MRS. B.

You see, my dear, how easily you have become acquainted with these modifications of insensible heat, which at first appeared so unintelligible. If, now, we were to reverse these changes, and condense the vapour into water, and the water into ice, the latent heat would re-appear entirely, in the form of free caloric.

EMILY.

Pray do let us see the effect of latent heat returning to its free state.

MRS. B.

For the purpose of showing this, we need simply conduct the vapour through this tube into this vessel of cold water, where it will part with its latent heat and return to its liquid form.

EMILY.

How rapidly the steam heats the water!

MRS. B.

That is because it does not merely impart its free caloric to the water, but likewise its latent heat. This method of heating liquids, has been turned to advantage, in several economical establishments. The steam-kitchens, which are getting into such general use, are upon the same principle. The steam is conveyed through a pipe in a similar manner, into the several vessels which contain the provisions to be dressed, where it communicates to them its latent caloric, and returns to the state of water. Count Rumford makes great use of this principle in many of his fire-places: his grand maxim is to avoid all unnecessary waste of caloric, for which purpose he confines the heat in such a manner, that not a particle of it shall unnecessarily escape; and while he economises the free caloric, he takes care also to turn the latent heat to advantage. It is thus that he is enabled to produce a degree of heat superior to that which is obtained in common fire-places, though he employs less fuel.

EMILY.

When the advantages of such contrivances are so clear and plain, I cannot understand why they are not universally used.

MRS. B.

A long time is always required before innovations, however useful, can be reconciled with the prejudices of the vulgar.

EMILY.

What a pity it is that there should be a prejudice against new inventions; how much more rapidly the world would improve, if such useful discoveries were immediately and universally adopted!

MRS. B.

I believe, my dear, that there are as many novelties attempted to be introduced, the adoption of which would be prejudicial to society, as there are of those which would be beneficial to it. The well-informed, though by no means exempt from error, have an unquestionable advantage over the illiterate, in judging what is likely or not to prove serviceable; and therefore we find the former more ready to adopt such discoveries as promise to be really advantageous, than the latter, who having no other test of the value of a novelty but time and experience, at first oppose its introduction. The well-informed, however, are frequently disappointed in their most sanguine expectations, and the prejudices of the vulgar, though they often retard the progress of knowledge, yet sometimes, it must be admitted, prevent the propagation of error. — But we are deviating from our subject.

We have converted steam into water, and are now to change water into ice, in order to render the latent heat sensible, as it escapes from the water on its becoming solid. For this purpose we must produce a degree of cold that will make water freeze.

CAROLINE.

That must be very difficult to accomplish in this warm room.

MRS. B.

Not so much as you think. There are certain chemical mixtures which produce a rapid change from the solid to the fluid state, or the reverse, in the substances combined, in consequence of which change latent heat is either extricated or absorbed.

EMILY.

I do not quite understand you.

MRS. B.

This snow and salt, which you see me mix together, are melting rapidly; heat, therefore, must be absorbed by the mixture, and cold produced.

CAROLINE.

It feels even colder than ice, and yet the snow is melted. This is very extraordinary.

MRS. B.

The cause of the intense cold of the mixture is to be attributed to the change from a solid to a fluid state. The union of the snow and salt produces a new arrangement of their particles, in consequence of which they become liquid; and the quantity of caloric, required to effect this change, is seized upon by the mixture wherever it can be obtained. This eagerness of the mixture for caloric, during its liquefaction, is such, that it converts part of its own free caloric into latent heat, and it is thus that its temperature is lowered.

EMILY.

Whatever you put in this mixture, therefore, would freeze?

MRS. B.

Yes; at least any fluid that is susceptible of freezing at that temperature. I have prepared this mixture of salt and snow for the purpose of freezing the water from which you are desirous of seeing the latent heat escape. I have put a thermometer in the glass of water that is to be frozen, in order that you may see how it cools.

CAROLINE.

The thermometer descends, but the heat which the water is now losing, is its *free*, not its *latent* heat.

MRS. B.

Certainly; it does not part with its latent heat till it changes its state and is converted into ice.

EMILY.

But here is a very extraordinary circumstance The thermometer is fallen below the freezing point, and yet the water is not frozen.

MRS. B.

That is always the case previous to the freezing of water when it is in a state of rest. Now it begins to congeal, and you may observe that the thermometer again rises to the freezing point.

CAROLINE.

It appears to me very strange that the thermometer should rise the very moment that the water freezes; for it seems to imply that the water was colder before it froze than when in the act of freezing.

MRS. B.

It is so; and after our long dissertation on this circumstance, I did not think it would appear so surprising to you. Reflect a little, and I think you will discover the reason of it.

CAROLINE.

It must be, no doubt, the extrications of latent heat, at the instant the water freezes, that raises the temperature.

MRS. B.

Certainly; and if you now examine the thermometer, you will find that its rise was but temporary, and lasted only during the disengagement of the latent heat — now that all the water is frozen it falls again, and will continue to fall till the ice and mixture are of an equal temperature.

EMILY.

And can you show us any experiments in which liquids, by being mixed, become solid, and disengage latent heat?

MRS. B.

I could show you several; but you are not yet sufficiently advanced to understand them well. I shall, however, try one, which will afford you a striking instance of the fact. The fluid which you see in this phial consists of a quantity of a certain salt called *muriat of lime*, dissolved in water. Now, if I pour into it a few drops of this other fluid, called *sulphuric acid*, the whole, or very nearly the whole, will be instantaneously converted into a solid mass.

EMILY.

How white it turns! I feel the latent heat escaping, for the bottle is warm, and the fluid is changed to a solid white substance like chalk!

CAROLINE.

This is, indeed, the most curious experiment we have seen yet. But pray what is that white vapour that ascends from the mixture?

MRS. B.

You are not yet enough of a chemist to understand that. — But take care, Caroline, do not approach too near it, for it has a very pungent smell. I shall show you another instance similar to that of the water, which you observed to become warmer as it froze. I have in this phial a solution of a salt called sulphat of soda or Glauber's salt, made very strong, and corked up when it was hot, and kept without agitation till it became cold, as you may feel the phial is. Now when I take out the cork and let the air fall upon it, (for being closed when boiling, there was a vacuum in the upper part) observe that the salt will suddenly crystallize. . . .

CAROLINE.

Surprising! how beautifully the needles of salt have shot through the whole phial!

MRS. B.

Yes, it is very striking — but pray do not forget the object of the experiment. Feel how warm the phial has become by the conversion of part of the liquid into a solid.

EMILY.

Quite warm I declare! this is a most curious experiment of the disengagement of latent heat.

MRS. B.

The slakeing of lime is another remarkable instance of the extrication of latent heat. Have you never observed how quick-lime smokes when water is poured upon it, and how much heat it produces?

CAROLINE.

Yes; but I do not understand what change of state takes place in the lime that occasions its giving out latent heat; for the quick-lime, which is solid, is (if I recollect right) reduced to powder, by this operation, and is, therefore, rather expanded than condensed.

MRS. B.

It is from the water, not the lime, that the latent heat is set free. The water incorporates with, and becomes solid in the lime; in consequence of which, the heat, which kept it in a liquid state, is disengaged, and escapes in a sensible form.

CAROLINE.

I always thought that the heat originated in the lime. It seems very strange that water, and cold water too, should contain so much heat.

EMILY.

After this extrication of caloric, the water must exist in a state of ice in the lime, since it parts with the heat which kept it liquid.

MRS. B.

It cannot properly be called ice, since ice implies a degree of cold, at least equal to the freezing point. Yet as water, in combining with lime, gives out more heat than in freezing, it must be in a state of still greater solidity in the lime, than it is in the form of ice; and you may have observed that *it* does not moisten or liquefy the lime in the smallest degree.

EMILY.

But, Mrs. B., the smoke that rises is white; if it was only pure caloric which escaped, we might feel, but could not see it.

MRS. B.

This white vapour is formed by some of the particles of lime, in a state of fine dust, which are carried off by the caloric.

EMILY.

In all changes of state, then, a body either absorbs or disengages latent heat?

MRS. B.

You cannot exactly say *absorbs latent heat*, as the heat becomes latent only on being confined in the body; but you may say, generally, that bodies, in passing from a solid to a liquid form, or from the liquid state to that of vapour, absorb heat; and that when the reverse takes place, heat is disengaged. *

EMILY.

We can now, I think, account for the ether boiling, and the water freezing in vacuo, at the same temperature. †

MRS. B.

Let me hear how you explain it.

EMILY.

The latent heat, which the water gave out in freezing, was immediately absorbed by the ether, during its conversion into vapour; and therefore, from a latent state in one liquid, it passed into a latent state in the other.

MRS. B.

But this only partly accounts for the result of the experiment; it remains to be explained why the

* This rule, if not universal, admits of very few exceptions.

† See page 102.

temperature of the ether, while in a state of ebullition, is brought down to the freezing temperature of the water. — It is because the ether, during its evaporation, reduces its own temperature, in the same proportion as that of the water, by converting its free caloric into latent heat: so that, though one liquid boils, and the other freezes, their temperatures remain in a state of equilibrium.

EMILY.

But why does not water, as well as ether, reduce its own temperature by evaporating?

MRS. B.

The fact is that it does, though much less rapidly than ether. Thus, for instance, you may often have observed, in the heat of summer, how much any particular spot may be cooled by watering, though the water used for that purpose be as warm as the air itself. Indeed so much cold may be produced by the mere evaporation of water, that the inhabitants of India, by availing themselves of the most favourable circumstances for this process which their warm climate can afford, namely, the cool of the night, and situations most exposed to the night breeze, succeed in causing water to freeze, though the temperature of the air be as high as 60 degrees. The water is put into shallow earthen trays, so as to expose an extensive

Fig. 1.
Voltaic Battery.

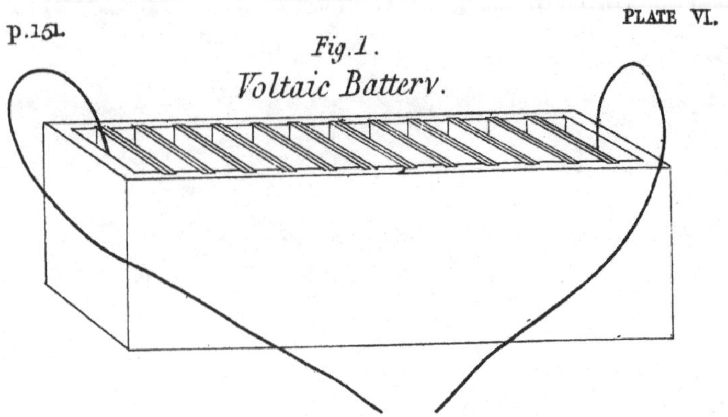

Fig. 2.

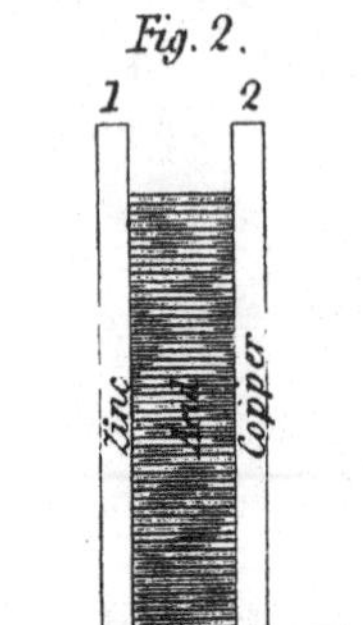

Fig. 4.

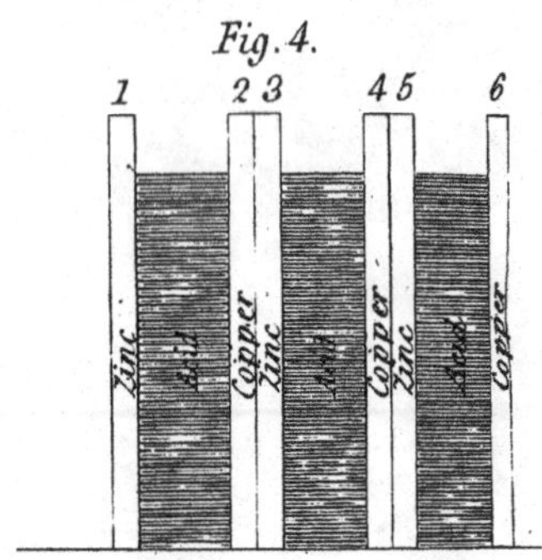

Fig. 3.
Electrical Machine.

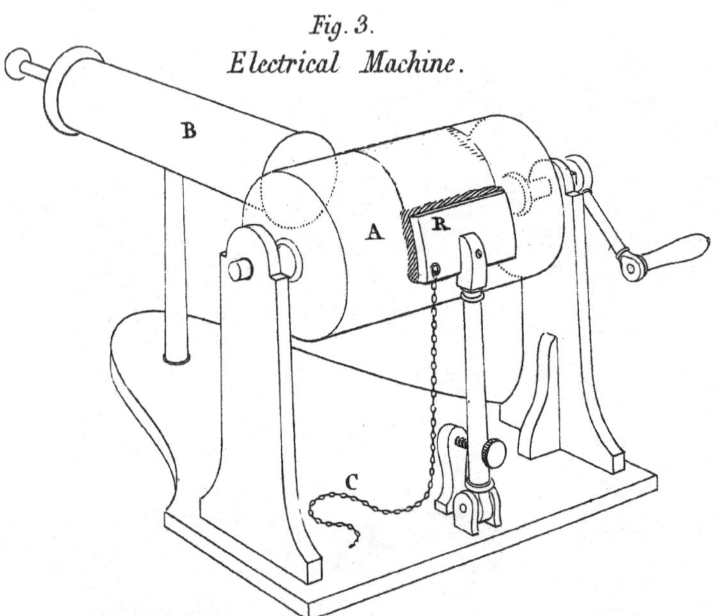

Fig. 3. A the Cylinder. _B the Conductor. _R the Rubber. _C the Chain.
Fig. 1. 2. & 4. Voltaic Batteries

Drawn by the Author

Published by Longman & Co. Octr 2. 1800.

surface to the process of evaporation, and in the morning, the water is found covered with a thin cake of ice, which is collected in sufficient quantity to be used for purposes of luxury.

CAROLINE.

How delicious it must be to drink liquids so cold in those tropical climates! But, Mrs. B., could we not try that experiment?

MRS. B.

If we were in the country, I have no doubt but that we should be able to freeze water, by the same means, and under similar circumstances. But we can do it immediately, upon a small scale, in this very room, in which the thermometer stands at 70 degrees. For this purpose we need only place some water in a little cup under the receiver of the air-pump (PLATE V. fig. 1.), and exhaust the air from it. What will be the consequence, Caroline?

CAROLINE.

Of course the water will evaporate more quickly, since there will no longer be any atmospheric pressure on its surface: but will this be sufficient to make the water freeze?

MRS. B.

Probably not, because the vapour will not be carried off fast enough; but this will be accomplished without difficulty if we introduce into the receiver (fig. 1.), in a saucer, or other large shallow vessel, some strong sulphuric acid, a substance which has a great attraction for water, whether in the form of vapour, or in the liquid state. This attraction is such that the acid will instantly absorb the moisture as it rises from the water, so as to make room for the formation of fresh vapour; this will of course hasten the process, and the cold produced from the rapid evaporation of the water, will, in a few minutes, be sufficient to freeze its surface. * We shall now exhaust the air from the receiver.

EMILY.

Thousands of small bubbles already rise through the water from the internal surface of the cup; what is the reason of this?

MRS. B.

These are bubbles of air which were partly attached to the vessel, and partly diffused in the water itself; and they expand and rise in consequence of the atmospheric pressure being removed.

* This experiment was first devised by Mr. Leslie, and has since been modified in a variety of forms.

CAROLINE.

See, Mrs. B.; the thermometer in the cup is sinking fast; it has already descended to 40 degrees!

EMILY.

The water seems now and then violently agitated on the surface, as if it was boiling; and yet the thermometer is descending fast!

MRS. B.

You may call it *boiling*, if you please, for this appearance is, as well as boiling, owing to the rapid formation of vapour; but here, as you have just observed, it takes place from the surface, for it is only when heat is applied to the bottom of the vessel that the vapour is formed there.—Now crystals of ice are actually shooting all over the surface of the water.

CAROLINE.

How beautiful it is! The surface is now entirely frozen—but the thermometer remains at 32 degrees.

MRS. B.

And so it will, conformably with our doctrine of latent heat, until the whole of the water is frozen; but it will then again begin to descend lower and lower, in consequence of the evaporation which goes on from the surface of the ice.

EMILY.

This is a most interesting experiment; but it would be still more striking if no sulphuric acid were required.

MRS. B.

I will show you a freezing instrument, contrived by Dr. Wollaston, upon the same principle as Mr. Leslie's experiment, by which water may be frozen by its own evaporation alone, without the assistance of sulphuric acid.

This tube, which, as you see (PLATE V. fig. 2.), is terminated at each extremity by a bulb, one of which is half full of water, is internally perfectly exhausted of air; the consequence of this is, that the water in the bulb is always much disposed to evaporate. This evaporation, however, does not proceed sufficiently fast to freeze the water; but if the empty ball be cooled by some artificial means, so as to condense quickly the vapour which rises from the water, the process may be thus so much promoted as to cause the water to freeze in the other ball. Dr. Wollaston has called this instrument *Cryophorus*.

CAROLINE.

So that cold seems to perform here the same part which the sulphuric acid acted in Mr. Leslie's experiment.

MRS. B.

Exactly so; but let us try the experiment.

EMILY.

How will you cool the instrument? You have neither ice nor snow.

MRS. B.

True: but we have other means of effecting this.* You recollect what an intense cold can be produced by the evaporation of ether in an exhausted receiver. We shall inclose the bulb in this little bag of fine flannel (fig. 3.), then soke it in ether, and introduce it into the receiver of the air-pump. (Fig. 5.) For this purpose we shall find it more convenient to use a cryophorus of this shape (fig. 4.), as its elongated bul') passes easily through a brass plate which closes the top of the receiver. If we now exhaust the receiver quickly, you will see, in less than a minute, the water freeze in the other bulb, out of the receiver.

EMILY.

The bulb already looks quite dim, and small drops of water are condensing on its surface.

* This mode of making the experiment was proposed, and the particulars detailed, by Dr. Marcet, in the 34th vol. of Nicholson's Journal, page 119.

CAROLINE.

And now crystals of ice shoot all over the water. This is, indeed, a very curious experiment!

MRS. B.

You will see, some other day, that, by a similar method, even quicksilver may be frozen. — But we cannot at present indulge in any further digression.

Having advanced so far on the subject of heat, I may now give you an account of the calorimeter, an instrument invented by Lavoisier, upon the principles just explained, for the purpose of estimating the specific heat of bodies. It consists of a vessel, the inner surface of which is lined with ice, so as to form a sort of hollow globe of ice, in the midst of which the body, whose specific heat is to be ascertained, is placed. The ice absorbs caloric from this body, till it has brought it down to the freezing point; this caloric converts into water a certain portion of the ice which runs out through an aperture at the bottom of the machine; and the quantity of ice changed to water is a test of the quantity of caloric which the body has given out in descending from a certain temperature to the freezing point.

CAROLINE.

In this apparatus, I suppose, the milk, chalk,

and lead, would melt different quantities of ice, in proportion to their different capacities for caloric?

MRS. B.

Certainly: and thence we are able to ascertain, with precision, their respective capacities for heat. But the calorimeter affords us no more idea of the absolute quantity of heat contained in a body, than the thermometer; for though by means of it we extricate both the free and combined caloric, yet we extricate them only to a certain degree, which is the freezing point; and we know not how much they contain of either below that point.

EMILY.

According to the theory of latent heat, it appears to me that the weather should be warm when it freezes, and cold in a thaw: for latent heat is liberated from every substance that it freezes, and such a large supply of heat must warm the atmosphere; whilst, during a thaw, that very quantity of free heat must be taken from the atmosphere, and return to a latent state in the bodies which it thaws.

MRS. B.

Your observation is very natural; but consider that in a frost the atmosphere is so much colder than the earth, that all the caloric which it takes from the freezing bodies is insufficient to raise its temperature above the freezing point; otherwise the frost must cease. But if the quantity of latent heat extricated does not destroy the frost, it serves to moderate the suddenness of the change of temperature of the atmosphere, at the commencement both of frost, and of a thaw. In the first instance, its extrication diminishes the severity of the cold; and, in the latter, its absorption moderates the warmth occasioned by a thaw: it even sometimes produces a discernible chill, at the breaking up of a frost.

CAROLINE.

But what are the general causes that produce those sudden changes in the weather, especially from hot to cold, which we often experience?

MRS. B.

This question would lead us into meteorological discussions, to which I am by no means competent: One circumstance, however, we can easily understand. When the air has passed over cold countries, it will probably arrive here at a temperature much below our own, and then it must absorb heat from every object it meets with, which will produce a general fall of temperature.

CAROLINE.

But pray, now that we know so much of the

effects of heat, will you inform us whether it is really a distinct body, or, as I have heard, a peculiar kind of motion produced in bodies?

MRS. B.

As I before told you, there is yet much uncertainty as to the nature of these subtle agents. But I am inclined to consider heat not as mere motion, but as a separate substance. Late experiments too appear to make it a compound body, consisting of the two electricities, and in our next conversation I shall inform you of the principal facts on which that opinion is founded.

CONVERSATION V.

ON THE CHEMICAL AGENCIES OF ELECTRICITY.

MRS. B.

Before we proceed further it will be necessary to give you some account of certain properties of electricity, which have of late years been discovered to have an essential connection with the phenomena of chemistry.

CAROLINE.

It is ELECTRICITY, if I recollect right, which comes next in our list of simple substances?

MRS. B.

I have placed electricity in that list, rather from the necessity of classing it somewhere, than from any conviction that it has a right to that situation, for we are as yet so ignorant of its intimate nature, that we are unable to determine, not only whether it is simple or compound, but whether it is in fact a material agent; or, as Sir H. Davy has hinted, whether it may not be merely a property inherent in matter. As, however, it is necessary to adopt some hypothesis for the explanation of the discoveries which this agent has enabled us to make, I have chosen the opinion, at present most prevalent, which supposes the existence of two kinds of electricity, distinguished by the names of *positive* and *negative* electricity.

CAROLINE.

Well, I must confess, I do not feel nearly so interested in a science in which so much uncertainty prevails, as in those which rest upon established principles; I never was fond of electricity, because, however beautiful and curious the phenomena it exhibits may be, the theories, by which they were explained, appeared to me so various, so obscure and inadequate, that I always remained dissatisfied. I was in hopes that the new discoveries in electricity had thrown so great a light on the subject, that every thing respecting it would now have been clearly explained.

MRS. B.

That is a point which we are yet far from having attained. But, in spite of the imperfection of our theories, you will be amply repaid by the importance and novelty of the subject. The number of new facts which have already been ascertained, and the immense prospect of discovery which has lately been opened to us, will, I hope, ultimately lead to a perfect elucidation of this branch of natural science; but at present you must be contented with studying the effects, and in some degree explaining the phenomena, without aspiring to a precise knowledge of the remote cause of electricity.

You have already obtained some notions of electricity: in our present conversation, therefore, I shall confine myself to that part of the science which is of late discovery, and is more particularly connected with chemistry.

It was a trifling and accidental circumstance which first gave rise to this new branch of physical science. Galvani, a professor of natural philosophy at Bologna, being engaged (about twenty years ago) in some experiments on muscular irritability, observed, that when a piece of metal was laid on the nerve of a frog, recently dead, whilst the limb supplied by that nerve rested upon some other metal, the limb suddenly moved, on a communication being made between the two pieces of metal.

EMILY.

How is this communication made?

MRS. B.

Either by bringing the two metals into contact, or by connecting them by means of a metallic conductor. But without subjecting a frog to any cruel experiments, I can easily make you sensible of this kind of electric action. Here is a piece of zinc, (one of the metals I mentioned in the list of elementary bodies) — put it *under* your tongue, and this piece of silver *upon* your tongue, and let both the metals project a little beyond the tip of the tongue — very well — now make the projecting parts of the metals touch each other, and you will instantly perceive a peculiar sensation.

EMILY.

Indeed I did, a singular taste, and I think a degree of heat: but I can hardly describe it.

MRS. B.

The action of these two pieces of metal on the tongue is, I believe, precisely similar to that made on the nerve of a frog. I shall not detain you by a detailed account of the theory by which Galvani attempted to account for this fact, as his explanation was soon overturned by subsequent experiments, which proved that *Galvanism* (the name this new power had obtained) was nothing more than electricity. Galvani supposed that the virtue of this new agent resided in the nerves of the frog, but Volta, who prosecuted this subject with much greater success, shewed that the phenomena did not depend on the organs of the frog, but upon the electrical agency of the metals, which is excited by the moisture of the animal, the organs of the frog being only a delicate test of the presence of electric influence.

CAROLINE.

I suppose, then, the saliva of the mouth answers the same purpose as the moisture of the frog, in exciting the electricity of the pieces of silver and zinc with which Emily tried the experiment on her tongue.

MRS. B.

Precisely. It does not appear, however, necessary that the fluid used for this purpose should be of an animal nature. Water, and acids very much diluted by water, are found to be the most effectual in promoting the developement of electricity in metals; and, accordingly, the original apparatus which Volta first constructed for this purpose, consisted of a pile or succession of plates of zinc and copper, each pair of which was connected by pieces of cloth or paper impregnated with water; and this instrument, from its original inconvenient structure and limited strength, has gradually arrived at its present state of power and improvement, such as is exhibited in the Voltaic battery. In this apparatus, a specimen of which you see before you (Plate VI. fig. 1.), the plates of zinc and copper are soldered together in pairs, each pair being placed at regular distances in wooden throughs and the interstices being filled with fluid.

CAROLINE.

Though you will not allow us to enquire into the precise cause of electricity, may we not ask in what manner the fluid acts on the metals so as to produce it?

MRS. B.

The action of the fluid on the metals, whether water or acid be used, is entirely of a chemical nature. But whether electricity is excited by this chemical action, or whether it is produced by the contact of the two metals, is a point upon which philosophers do not vet perfectly agree.

EMILY.

But can the mere contact of two metals, without any intervening fluid, produce electricity?

MRS. B.

Yes, if they are afterwards separated. It is an established fact, that when two metals are put in contact, and afterwards separated, that which has the strongest attraction for oxygen exhibits signs of positive, the other of negative electricity.

CAROLINE.

It seems then but reasonable to infer that the

power of the Voltaic battery should arise from the contact of the plates of zinc and copper.

MRS. B.

It is upon this principle that Volta and Sir H. Davy explain the phenomena of the pile; but notwithstanding these two great authorities, many philosophers entertain doubts on the truth of this theory. The principal difficulty which occurs in explaining the phenomena of the Voltaic battery on this principle, is, that two such plates show no signs of different states of electricity whilst in contact, but only on being separated after contact. Now in the Voltaic battery, those plates that are in contact always continue so, being soldered together: and they cannot therefore receive a succession of charges. Besides, if we consider the mere disturbance of the balance of electricity by the contact of the plates, as the sole cause of the production of Voltaic electricity, it remains to be explained how this disturbed balance becomes an inexhaustible source of electrical energy, capable of pouring forth a constant and copious supply of electrical fluid, though without any means of replenishing itself from other sources. This subject, it must be owned, is involved in too much obscurity to enable us to speak very decidedly in favour of any theory. But, in order to avoid perplexing you with different explanations,

I shall confine myself to one which appears to me to be least encumbered with difficulties, and most likely to accord with truth.*

This theory supposes the electricity to be excited by the chemical action of the acid on the zinc; but you are yet such novices in chemistry, that I think it will be necessary to give you some previous explanation of the nature of this action.

All metals have a strong attraction for oxygen, and this element is found in great abundance both in water and in acids. The action of the diluted acid on the zinc consists therefore in its oxygen combining with it, and dissolving its surface.

CAROLINE.

In the same manner I suppose as we saw an acid dissolve copper?

MRS. B.

Yes; but in the Voltaic battery the diluted acid is not strong enough to produce so complete

* This mode of explaining the phenomena of the Voltaic pile is called the *chemical theory* of electricity, because it ascribes the cause of these phenomena to certain chemical changes which take place during their appearance. In the preceding edition of this work, the same theory was presented in a more elaborate, but less easy form than it is in this. The mode of viewing the subject which is here sketched was long since suggested by Dr. Bostock, of whose theory, however, this is by no means to be considered as a complete statement.

an effect; it acts only on the surface of the zinc, to which it yields its oxygen, forming upon it a film or crust, which is a compound of the oxygen and the metal.

EMILY.

Since there is so strong a chemical attraction between oxygen and metals, I suppose they are naturally in different states of electricity?

MRS. B.

Yes; it appears that all metals are united with the positive, and that oxygen is the grand source of the negative electricity.

CAROLINE.

Does not then the acid act on the plates of copper, as well as on those of zinc?

MRS. B.

No; for though copper has an affinity for oxygen, it is less strong than that of zinc; and therefore the energy of the acid is only exerted upon the zinc.

It will be best, I believe, in order to render the action of the Voltaic battery more intelligible, to confine our attention at first to the effect produced on two plates only. (Plate VI. fig. 2.)

If a plate of zinc be placed opposite to one of copper, or any other metal less attractive of oxygen, and the space between them (suppose of half an inch in thickness), be filled with an acid or any fluid capable of oxydating the zinc, the oxydated surface will have its capacity for electricity diminished, so that a quantity of electricity will be evolved from that surface. This electricity will be received by the contiguous fluid, by which it will be transmitted to the opposite metallic surface, the copper, which is not oxydated, and is therefore disposed to receive it; so that the copper plate will thus become positive, whilst the zinc plate will be in the negative state.

This evolution of electrical fluid however will be very limited; for as these two plates admit of but very little accumulation of electricity, and are supposed to have no communication with other bodies, the action of the acid, and further developement of electricity, will be immediately stopped.

EMILY.

This action, I suppose, can no more continue to go on, than that of a common electrical machine, which is not allowed to communicate with other bodies?

MRS. B.

Precisely; the common electrical machine, when excited by the friction of the rubber, gives out both the positive and negative electricities.— (PLATE VI. Fig. 3.) The positive, by the rotation of the glass cylinder, is conveyed into the conductor, whilst the negative goes into the rubber. But unless there is a communication made between the rubber and the ground, but a very inconsiderable quantity of electricity can be excited; for the rubber, like the plates of the battery, has too small a capacity to admit of an accumulation of electricity. Unless therefore the electricity can pass out of the rubber, it will not continue to go into it, and consequently no additional accumulation will take place. Now as one kind of electricity cannot be given out without the other, the developement of the positive electricity is stopped as well as that of the negative, and the conductor therefore cannot receive a succession of charges.

CAROLINE.

But does not the conductor, as well as the rubber, require a communication with the earth, in order to get rid of its electricity?

MRS. B.

No; for it is susceptible of receiving and containing a considerable quantity of electricity, as it is much larger than the rubber, and therefore has a greater capacity; and this continued accumulation of electricity in the conductor is what is called a charge.

EMILY.

But when an electrical machine is furnished with two conductors to receive the two electricities, I suppose no communication with the earth is required?

MRS. B.

Certainly not, until the two are fully charged; for the two conductors will receive equal quantities of electricity.

CAROLINE.

I thought the use of the chain had been to convey the electricity *from* the ground into the machine?

MRS. B.

That was the idea of Dr. Franklin, who supposed that there was but one kind of electricity, and who, by the terms positive and negative (which he first introduced), meant only different quantities of the same kind of electricity. The chain was in that case supposed to convey electricity *from* the ground through the rubber into the conductor. But as we have adopted the hypothesis of two electricities, we must consider the chain as a vehicle to conduct the negative electricity into the earth.

EMILY.

And are both kinds of electricity produced whenever electricity is excited?

MRS. B.

Yes, invariably. If you rub a tube of glass with a woollen cloth, the glass becomes positive, and the cloth negative. If, on the contrary, you excite a stick of sealing-wax by the same means, it is the rubber which becomes positive, and the wax negative.

But with regard to the Voltaic battery, in order that the acid may act freely on the zinc, and the two electricities be given out without interruption, some method must be devised, by which the plates may part with their electricities as fast as they receive them. — Can you think of any means by which this might be effected?

EMILY.

Would not two chains or wires, suspended from either plate to the ground, conduct the electricities into the earth, and thus answer the purpose?

MRS. B.

It would answer the purpose of carrying off the electricity, I admit; but recollect, that though it is necessary to find a vent for the electricity, yet we must not lose it, since it is the power which we are endeavouring to obtain. Instead, therefore, of conducting it into the ground, let us make the wires, from either plate, meet: the two electricities will thus be brought together, and will combine and neutralize each other; and as long as this communication continues, the two plates having a vent for their respective electricities, the action of the acid will go on freely and uninterruptedly.

EMILY.

That is very clear, so far as two plates only are concerned; but I cannot say I understand how the energy of the succession of plates, or rather pairs of plates, of which the Galvanic trough is composed, is propagated and accumulated throughout a battery?

MRS. B.

In order to shew you how the intensity of the electricity is increased by increasing the number of plates, we will examine the action of four plates; if you understand these, you will readily comprehend that of any number whatever. In this figure (PLATE VI. Fig. 4.), you will observe that the two central plates are united; they are soldered together, (as we observed in describing the Voltaic trough,) so as to form but one plate which offers two different surfaces, the one of copper, the other of zinc.

Now you recollect that, in explaining the action of two plates, we supposed that a quantity of electricity was evolved from the surface of the first zinc plate, in consequence of the action of the acid, and was conveyed by the interposed fluid to the copper plate, No. 2, which thus became positive. This copper plate communicates its electricity to the contiguous zinc plate, No. 3, in which, consequently, some accumulation of electricity takes place. When, therefore, the fluid in the next cell acts upon the zinc plate, electricity is extricated from it in larger quantity, and in a more concentrated form, than before. This concentrated electricity is again conveyed by the fluid to the next pair of plates, No. 4 and 5, when it is farther increased by the action of the fluid in the third cell, and so on, to any number of plates of which the battery may consist; so that the electrical energy will continue to accumulate in proportion to the number of double plates, the first zinc plate of the series being the most negative, and the last copper plate the most positive.

CAROLINE.

But does the battery become more and more strongly charged, merely by being allowed to stand undisturbed?

MRS. B.

No, for the action will soon stop, as was explained before, unless a vent be given to the accumulated electricities. This is easily done, however, by establishing a communication by means of the wires (Fig. 1.), between the two ends of the battery: these being brought into contact, the two electricities meet and neutralize each other, producing the shock and other effects of electricity; and the action goes on with renewed energy, being no longer obstructed by the accumulation of the two electricities which impeded its progress.

EMILY.

Is it the union of the two electricities which produces the electric spark?

MRS. B.

Yes; and it is, I believe, this circumstance which gave rise to Sir H. Davy's opinion that caloric may be a compound of the two electricities.

CAROLINE.

Yet surely caloric is very different from the electrical spark?

MRS. B.

The difference may consist probably only in intensity: for the heat of the electric spark is considerably more intense, though confined to a very minute spot, than any heat we can produce by other means.

EMILY.

Is it quite certain that the electricity of the Voltaic battery is precisely of the same nature as that of the common electrical machine?

MRS. B.

Undoubtedly; the shock given to the human body, the spark, the circumstance of the same substances which are conductors of the one being also conductors of the other, and of those bodies, such as glass and sealing-wax, which are non-conductors of the one, being also non-conductors of the other, are striking proofs of it. Besides, Sir H. Davy has shewn in his Lectures, that a Leyden jar, and a common electric battery, can be charged with electricity obtained from a Voltaic battery, the effect produced being perfectly similar to that obtained by a common machine.

Dr. Wollaston has likewise proved that similar chemical decompositions are effected by the electric machine and by the Voltaic battery; and has made other experiments which render it highly probable, that the origin of both electricities is essentially the same, as they show that the rubber of the common electrical machine, like the zinc in the Voltaic battery, produces the two electricities by combining with oxygen.

CAROLINE.

But I do not see whence the rubber obtains oxygen, for there is neither acid nor water used in the common machine, and I always understood that the electricity was excited by the friction.

MRS. B.

It appears that by friction the rubber obtains oxygen from the atmosphere, which is partly composed of that element. The oxygen combines with the amalgam of the rubber, which is of a metallic nature, much in the same way as the oxygen of the acid combines with the zinc in the Voltaic battery, and it is thus that the two electricities are disengaged.

CAROLINE.

But, if the electricities of both machines are similar, why not use the common machine for chemical decompositions?

MRS. B.

Though its effects are similar to those of the Voltaic battery, they are incomparably weaker. Indeed Dr. Wollaston, in using it for chemical decompositions, was obliged to act upon the most minute quantities of matter, and though the result was satisfactory in proving the similarity of its effects to those of the Voltaic battery, these effects were too small in extent to be in any considerable degree applicable to chemical decomposition.

CAROLINE.

How terrible, then, the shock must be from a Voltaic battery, since it is so much more powerful than an electrical machine!

MRS. B.

It is not nearly so formidable as you think; at least it is by no means proportional to the chemical effect. The great superiority of the Voltaic battery consists in the large *quantity* of electricity that passes; but in regard to the *rapidity* or *intensity* of the charge, it is greatly surpassed by the common electrical machine. It would seem that the shock or sensation depends chiefly upon the intensity; whilst, on the contrary, for chemical purposes, it is quantity which is required. In the Voltaic battery, the electricity, though copious, is so weak as not to be able to force its way through the fluid which separates the plates, whilst that of a common machine will pass through any space of water.

CAROLINE.

Would not it be possible to increase the intensity of the Voltaic battery till it should equal that of the common machine?

MRS. B.

It can actually be increased till it imitates a weak electrical machine, so as to produce a visible spark when accumulated in a Leyden jar. But it can never be raised sufficiently to pass through any considerable extent of air, because of the ready communication through the fluids employed.

By increasing the number of plates of a battery,

you increase its *intensity*, whilst, by enlarging the dimensions of the plates, you augment its *quantity*; and, as the superiority of the battery over the common machine consists entirely in the quantity of electricity produced, it was at first supposed that it was the size, rather than the number of plates that was essential to the augmentation of power. It was, however, found upon trial, that the quantity of electricity produced by the Voltaic battery, even when of a very moderate size, was sufficiently copious, and that the chief advantage in this apparatus was obtained by increasing the intensity, which, however, still falls very short of that of the common machine.

I should not omit to mention, that a very splendid, and, at the same time, most powerful battery, was, a few years ago, constructed under the direction of Sir H. Davy, which he repeatedly exhibited in his course of electro-chemical lectures. It consists of two thousand double plates of zinc and copper, of six square inches in dimensions, arranged in troughs of Wedgwood-ware, each of which contains twenty of these plates. The troughs are furnished with a contrivance for lifting the plates out of them in a very convenient and expeditious manner.*

* A model of this mode of construction is exhibited in PLATE XII. Fig. 1.

CAROLINE.

Well, now that we understand the nature of the action of the Votaic battery, I long to hear an account of the discoveries to which it has given rise.

MRS. B.

You must restrain your impatience, my dear, for I cannot with any propriety introduce the subject of these discoveries till we come to them in the regular course of our studies. But, as almost every substance in nature has already been exposed to the influence of the Voltaic battery, we shall very soon have occasion to notice its effects.

PLATE VII.

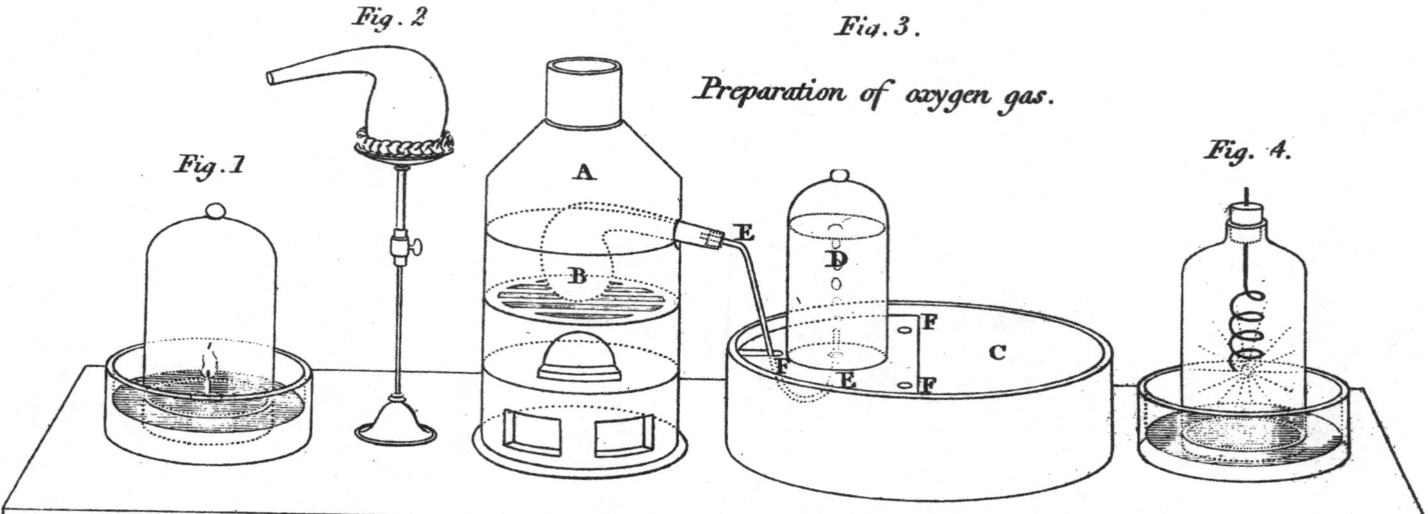

Fig. 1, Combustion of a taper under a receiver.—Fig. 2. A Retort on a stand.—Fig. 3. A Furnace. B Earthen Retort in the furnace. C Water bath. D Receiver. E.E Tube conveying the gas from the Retort through the water into the Receiver. F.F.F Shelf perforated on which the Receiver stands. Fig. 4. Combustion of iron wire in oxygen gas.

Drawn by the Author.

Engraved by Lowry

Published by Longman & Co. Octr. 2nd. 1809.

CONVERSATION VI.

ON OXYGEN AND NITROGEN.

MRS. B.

TO-DAY we shall examine the chemical properties of the ATMOSPHERE.

CAROLINE.

I thought that we were first to learn the nature of OXYGEN, which come next in our table of simple bodies?

MRS. B.

And so you shall; the atmosphere being composed of two principles, OXYGEN and NITROGEN, we shall proceed to analyse it, and consider its component parts separately.

EMILY.

I always thought that the atmosphere had been a very complicated fluid, composed of all the variety of exhalations from the earth.

MRS. B.

Such substances may be considered rather as heterogeneous and accidental, than as forming any of its component parts; and the proportion they bear to the whole mass is quite inconsiderable.

Atmospherical air is composed of two gasses, known by the names of oxygen gas and nitrogen or azotic gas.

EMILY.

Pray what is a gas?

MRS. B.

The name of gas is given to any fluid capable of existing constantly in an aeriform state, under the pressure and at the temperature of the atmosphere.

CAROLINE.

Is not water, or any other substance, when evaporated by heat, called gas?

MRS. B.

No, my dear; vapour is, indeed, an elastic fluid, and bears a strong resemblance to a gas; there are, however, several points in which they essentially differ, and by which you may always distinguish them. Steam, or vapour, owes its elasticity merely to a high temperature, which is equal to that of boiling water. And it differs from boiling water only by being united with more caloric, which, as we before explained, is in a latent state. When steam is cooled, it instantly returns to the form of water; but air, or gas, has never yet been rendered liquid or solid by any degree of cold.

EMILY.

But does not gas, as well as vapour, owe its elasticity to caloric?

MRS. B.

It was the prevailing opinion; and the difference of gas or vapour was thought to depend on the different manner in which caloric was united with the basis of these two kinds of elastic fluids. In vapour, it was considered as in a latent state; in gas, it was said to be chemically combined. But the late researches of Sir H. Davy have given rise to a new theory respecting gasses; and there is now reason to believe that these bodies owe their permanently elastic state, not solely to caloric, but likewise to the prevalence of either the one or the other of the two electricities.

EMILY.

When you speak, then, of the simple bodies oxygen and nitrogen, you mean to express those substances which are the basis of the two gasses?

MRS. B.

Yes, in strict propriety, for they can properly be called gasses only when brought to an aeriform state.

CAROLINE.

In what proportions are they combined in the atmosphere?

MRS. B.

The oxygen gas constitutes a little more than one-fifth, and the nitrogen gas a little less than four-fifths. When separated, they are found to possess qualities totally different from each other. For oxygen gas is essential both to respiration and combustion, while neither of these processes can be performed in nitrogen gas.

CAROLINE.

But if nitrogen gas is unfit for respiration, how does it happen that the large proportion of it which enters into the composition of the atmosphere is not a great impediment to breathing?

MRS. B.

We should breathe more freely than our lungs could bear, if we respired oxygen gas alone. The nitrogen is no impediment to respiration, and probably, on the contrary, answers some useful purpose, though we do not know in what manner it acts in that process.

EMILY.

And by what means can the two gasses, which compose the atmospheric air, be separated?

MRS. B.

There are many ways of analysing the atmosphere: the two gasses may be separated first by combustion.

EMILY.

You surprise me! how is it possible that combustion should separate them?

MRS. B.

I should previously remind you that oxygen is supposed to be the only simple body naturally combined with negative electricity. In all the other elements the positive electricity prevails, and they have consequently, all of them, an attraction for oxygen. *

CAROLINE.

Oxygen the only negatively electrified body! that surprises me extremely; how then are the combinations of the other bodies performed, if, according to your explanation of chemical attraction, bodies are supposed only to combine in virtue of their opposite states of electricity?

* If chlorine or oxymuriatic gas be a simple body, according to Sir H. Davy's view of the subject, it must be considered as an exception to this statement; but this subject cannot be discussed till the properties and nature of chlorine come under examination.

MRS. B.

Observe that I said, that oxygen was the only *simple* body, naturally negative. Compound bodies, in which oxygen prevails over the other component parts, are also negative, but their negative energy is greater or less in proportion as the oxygen predominates. Those compounds into which oxygen enters in less proportion than the other constituents, are positive, but their positive energy is diminished in proportion to the quantity of oxygen which enters into their composition.

All bodies, therefore, that are not already combined with oxygen, will attract it, and, under certain circumstances, will absorb it from the atmosphere, in which case the nitrogen gas will remain alone, and may thus be obtained in its separate state.

CAROLINE.

I do not understand how a gas can be absorbed?

MRS. B.

It is only the oxygen, or basis of the gas, which is absorbed; and the two electricities escaping, that is to say, the negative from the oxygen, the positive from the burning body, unite and produce caloric.

EMILY.

And what becomes of this caloric?

MRS. B.

We shall make this piece of dry wood attract oxygen from the atmosphere, and you will see what becomes of the caloric.

CAROLINE.

You are joking, Mrs. B—; you do not mean to decompose the atmosphere with a piece of dry stick?

MRS. B.

Not the whole body of the atmosphere, certainly; but if we can make this piece of wood attract any quantity of oxygen from it, a proportional quantity of atmospherical air will be decomposed.

CAROLINE.

If wood has so strong an attraction for oxygen, why does it not decompose the atmosphere spontaneously?

MRS. B.

It is found by experience, that an elevation of temperature is required for the commencement of the union of the oxygen and the wood.

This elevation of temperature was formerly thought to be necessary, in order to diminish the cohesive attraction of the wood, and enable the oxygen to penetrate and combine with it more readily. But since the introduction of the new theory of chemical combination, another cause has been assigned, and it is now supposed that the high temperature, by exalting the electrical energies of bodies, and consequently their force of attraction, facilitates their combination.

EMILY.

If it is true, that caloric is composed of the two electricities, an elevation of temperature must necessarily augment the electric energies of bodies.

MRS. B.

I doubt whether that would be a necessary consequence; for, admitting this composition of caloric, it is only by its being decomposed that electricity can be produced. Sir H. Davy, however, in his numerous experiments, has found it to be an almost invariable rule that the electrical energies of bodies are increased by elevation of temperature.

What means then shall we employ to raise the temperature of the wood, so as to enable it to attract oxygen from the atmosphere?

CAROLINE.

Holding it near the fire, I should think, would answer the purpose.

MRS. B.

It may, provided you hold it sufficiently close

to the fire; for a very considerable elevation of temperature is required.

CAROLINE.

It has actually taken fire, and yet I did not let it touch the coals, but I held it so very close that I suppose it caught fire merely from the intensity of the heat.

MRS. B.

Or you might say, in other words, that the caloric which the wood imbibed, so much elevated its temperature, and exalted its electric energy, as to enable it to attract oxygen very rapidly from the atmosphere.

EMILY.

Does the wood absorb oxygen while it is burning?

MRS. B.

Yes, and the heat and light are produced by the union of the two electricities which are set at liberty, in consequence of the oxygen combining with the wood.

CAROLINE.

You astonish me! the heat of a burning body proceeds then as much from the atmosphere as from the body itself?

MRS. B.

It was supposed that the caloric, given out

during combustion, proceeded entirely, or nearly so, from the decomposition of the oxygen gas; but, according to Sir H. Davy's new view of the subject, both the oxygen gas, and the combustible body, concur in supplying the heat and light, by the union of their opposite electricities.

EMILY.

I have not yet met with any thing in chemistry that has surprised or delighted me so much as this explanation of combustion. I was at first wondering what connection there could be between the affinity of a body for oxygen and its combustibility; but I think I understand it now perfectly.

MRS. B.

Combustion then, you see, is nothing more than the rapid combination of a body with oxygen, attended by the disengagement of light and heat.

EMILY.

But are there no combustible bodies whose attraction for oxygen is so strong, that they will combine with it, without the application of heat?

CAROLINE.

That cannot be; otherwise we should see bodies burning spontaneously.

MRS. B.

But there are some instances of this kind, such as phosphorus, potassium, and some compound bodies, which I shall hereafter make you acquainted with. These bodies, however, are prepared by art, for in general, all the combustions that could occur spontaneously, at the temperature of the atmosphere, have already taken place; therefore new combustions cannot happen without the temperature of the body being raised. Some bodies, however, will burn at a much lower temperature than others.

CAROLINE.

But the common way of burning a body is not merely to approach it to one already on fire, but rather to put the one in actual contact with the other, as when I burn this piece of paper by holding it in the flame of the fire.

MRS. B.

The closer it is in contact with the source of caloric, the sooner will its temperature be raised to the degree necessary for it to burn. If you hold it near the fire, the same effect will be produced; but more time will be required, as you found to be the case with the piece of stick.

EMILY.

But why is it not necessary to continue applying caloric throughout the process of combustion, in order to keep up the electric energy of the wood, which is required to enable it to combine with the oxygen?

MRS. B.

The caloric which is gradually produced by the two electricities during combustion, keeps up the temperature of the burning body; so that when once combustion has begun, no further application of caloric is required.

CAROLINE.

Since I have learnt this wonderful theory of combustion, I cannot take my eyes from the fire; and I can scarcely conceive that the heat and light, which I always supposed to proceed entirely from the coals, are really produced as much by the atmosphere.

EMILY.

When you blow the fire, you increase the combustion, I suppose, by supplying the coals with a greater quantity of oxygen gas?

MRS. B.

Certainly; but of course no blowing will produce combustion, unless the temperature of the coals be first raised. A single spark, however, is sometimes sufficient to produce that effect; for as I said before, when once combustion has commenced, the caloric disengaged is sufficient to elevate the temperature of the rest of the body, provided that there be a free access of oxygen. It however sometimes happens that if a fire be ill made, it will be extinguished before all the fuel is consumed, from the very circumstance of the combustion being so slow that the caloric disengaged is insufficient to keep up the temperature of the fuel. You must recollect that there are three things required in order to produce combustion; a combustible body, oxygen, and a temperature at which the one will combine with the other.

EMILY.

You said that combustion was one method of decomposing the atmosphere, and obtaining the nitrogen gas in its simple state; but how do you secure this gas, and prevent it from mixing with the rest of the atmosphere?

MRS. B.

It is necessary for this purpose to burn the body within a close vessel, which is easily done.—We shall introduce a small lighted taper (Plate VII. Fig. 1.) under this glass receiver, which stands in a bason over water, to prevent all communication with the external air.

CAROLINE.

How dim the light burns already!—It is now extinguished.

MRS. B.

Can you tell us why it is extinguished?

CAROLINE.

Let me consider.—The receiver was full of atmospherical air; the taper, in burning within it, must have combined with the oxygen contained in that air, and the caloric that was disengaged produced the light of the taper. But when the whole of the oxygen was absorbed, the whole of its electricity was disengaged; consequently no more caloric could be produced, the taper ceased to burn, and the flame was extinguished.

MRS. B.

Your explanation is perfectly correct.

EMILY.

The two constituents of the oxygen gas being thus disposed of, what remains under the receiver must be pure nitrogen gas?

MRS. B.

There are some circumstances which prevent the nitrogen gas, thus obtained, from being perfectly pure; but we may easily try whether the oxygen has disappeared, by putting another lighted taper under it.—You see how instantaneously the flame is extinguished, for want of oxygen to supply the negative electricity required for the formation of caloric; and were you to put an animal under the receiver, it would immediately be suffocated. But that is an experiment which I do not think your curiosity will tempt you to try.

EMILY.

Certainly not.—But look, Mrs. B., the receiver is full of a thick white smoke. Is that nitrogen gas?

MRS. B.

No, my dear; nitrogen gas is perfectly transparent and invisible, like common air. This cloudiness proceeds from a variety of exhalations, which arise from the burning taper, and the nature of which you cannot yet understand.

CAROLINE.

The water within the receiver has now risen a little above its level in the bason. What is the reason of this?

MRS. B.

With a moment's reflection, I dare say, you would have explained it yourself. The water rises in consequence of the oxygen gas within it having been destroyed, or rather decomposed, by the combustion of the taper.

CAROLINE.

Then why did not the water rise immediately when the oxygen gas was destroyed?

MRS. B.

Because the heat of the taper, whilst burning, produced a dilatation of the air in the vessel, which at first counteracted this effect.

Another means of decomposing the atmosphere is the *oxygenation* of certain metals. This process is very analogous to combustion; it is, indeed, only a more general term to express the combination of a body with oxygen.

CAROLINE.

In what respect, then, does it differ from combustion?

MRS. B.

The combination of oxygen in combustion is always accompanied by a disengagement of light and heat; whilst this circumstance is not a necessary consequence of simple oxygenation.

CAROLINE.

But how can a body absorb oxygen without the combination of the two electricities which produce caloric?

MRS. B.

Oxygen does not always present itself in a gâseous state; it is a constituent part of a vast number of bodies, both solid and liquid, in which it exists in a much denser state than in the atmosphere; and from these bodies it may be obtained without much disengagement of caloric. It may likewise, in some cases, be absorbed from the atmosphere without any sensible production of light and heat; for, if the process be slow, the caloric is disengaged in such small quantities, and so gradually, that it is not capable of producing either light or heat. In this case the absorption of oxygen is called *oxygenation* or *oxydation*, instead of *combustion*, as the production of sensible light and heat is essential to the latter.

EMILY.

I wonder that metals can unite with oxygen; for, as they are so dense, their attraction of aggregation must be very great; and I should have thought that oxygen could never have penetrated such bodies.

MRS. B.

Their strong attraction for oxygen counterbalances this obstacle. Most metals, however, require to be made red-hot before they are capable of attracting oxygen in any considerable quantity.

By this combination they lose most of their metallic properties, and fall into a kind of powder, formerly called *calx*, but now much more properly termed an *oxyd*; thus we have *oxyd of lead*, *oxyd of iron*, &c.

EMILY.

And in the Voltaic battery, it is, I suppose, an oxyd of zinc, that is formed by the union of the oxygen with that metal?

MRS. B.

Yes, it is.

CAROLINE.

The word oxyd, then, simply means a metal combined with oxygen?

MRS. B.

Yes; but the term is not confined to metals, though chiefly applied to them. Any body whatever, that has combined with a certain quantity of oxygen, either by means of oxydation or combustion, is called an *oxyd*, and is said to be *oxydated* or *oxygenated*.

EMILY.

Metals, when converted into oxyds, become, I suppose, negative?

MRS. B.

Not in general; because in most oxyds the positive energy of the metal more than counterbalances the native energy of the oxygen with which it combines.

This black powder is an oxyd of manganese, a metal which has so strong an affinity for oxygen, that it attracts that substance from the atmosphere at any known temperature: it is therefore never found in its metallic form, but always in that of an oxyd, in which state, you see, it has very little of the appearance of a metal. It is now heavier than it was before oxydation, in consequence of the additional weight of the oxygen with which it has combined.

CAROLINE.

I am very glad to hear that; for I confess I could not help having some doubts whether oxygen was really a substance, as it is not to be obtained in a simple and palpable state; but its weight is, I think, a decisive proof of its being a real body.

MRS. B.

It is easy to estimate its weight, by separating it from the manganese, and finding how much the latter has lost.

EMILY.

But if you can take the oxygen from the metal, shall we not then have it in its palpable simple state?

MRS. B.

No; for I can only separate the oxygen from

the manganese, by presenting to it some other body, for which it has a greater affinity than for the manganese. Caloric affording the two electricities is decomposed, and one of them uniting with the oxygen, restores it to the aëriform state.

EMILY.

But you said just now, that manganese would attract oxygen from the atmosphere in which it is combined with the negative electricity; how, therefore, can the oxygen have a superior affinity for that electricity, since it abandons it to combine with the manganese?

MRS. B.

I give you credit for this objection, Emily; and the only answer I can make to it is, that the mutual affinities of metals for oxygen, and of oxygen for electricity, vary at different temperatures; a certain degree of heat will, therefore, dispose a metal to combine with oxygen, whilst, on the contrary, the former will be compelled to part with the latter, when the temperature is further increased. I have put some oxyd of manganese into a retort, which is an earthen vessel with a bent neck, such as you see here. (Plate VII. Fig. 2.) — The retort containing the manganese you cannot see, as I have enclosed it in this furnace, where it is now red-hot. But, in order to make you sensible of the escape of the gas, which is itself invisible, I have connected the neck of the retort with this bent tube, the extremity of which is immersed in this vessel of water. (PLATE VII. Fig. 3.) — Do you see the bubbles of air rise through the water?

CAROLINE.

Perfectly. This, then, is pure oxygen gas; what a pity it should be lost! Could you not preserve it?

MRS. B.

We shall collect it in this receiver. — For this purpose, you observe, I first fill it with water, in order to exclude the atmospherical air; and then place it over the bubbles that issue from the retort, so as to make them rise through the water to the upper part of the receiver.

EMILY.

The bubbles of oxygen gas rise, I suppose, from their specific levity?

MRS. B.

Yes; for though oxygen forms rather a heavy gas, it is light compared to water. You see how it gradually displaces the water from the receiver. It is now full of gas, and I may leave it inverted in water on this shelf, where I can keep the gas as long as I choose, for future experiments. This apparatus (which is indispensable in all experiments in which gases are concerned) is called a water-bath.

CAROLINE.

It is a very clever contrivance, indeed; equally simple and useful. How convenient the shelf is for the receiver to rest upon under water, and the holes in it for the gas to pass into the receiver! I long to make some experiments with this apparatus.

MRS. B.

I shall try your skill that way, when you have a little more experience. I am now going to show you an experiment, which proves, in a very striking manner, how essential oxygen is to combustion. You will see that iron itself will burn in this gas, in the most rapid and brilliant manner.

CAROLINE.

Really! I did not know that it was possible to burn iron.

EMILY.

Iron is a simple body, and you know, Caroline, that all simple bodies are naturally positive, and therefore must have an affinity for oxygen.

MRS. B.

Iron will, however, not burn in atmospherical

air without a very great elevation of temperature; but it is eminently combustible in pure oxygen gas; and what will surprise you still more, it can be set on fire without any considerable rise of temperature. You see this spiral irón wire — I fasten it at one end to this cork, which is made to fit an opening at the top of the glass-receiver. (PLATE VII. Fig. 4.)

EMILY.

I see the opening in the receiver; but it is carefully closed by a ground glass-stopper.

MRS. B.

That is in order to prevent the gas from escaping; but I shall take out the stopper, and put in the cork, to which the wire hangs. — Now I mean to burn this wire in the oxygen gas, but I must fix a small piece of lighted tinder to the extremity of it, in order to give the first impulse to combustion; for, however powerful oxygen is in promoting combustion, you must recollect that it cannot take place without some elevation of temperature. I shall now introduce the wire into the receiver, by quickly changing the stoppers.

CAROLINE.

Is there no danger of the gas escaping while you change the stoppers?

MRS. B.

Oxygen gas is a little heavier than atmospherical air, therefore it will not mix with it very rapidly; and, if I do not leave the opening uncovered, we shall not lose any——

CAROLINE.

Oh, what a brilliant and beautiful flame!

EMILY.

It is as white and dazzling as the sun! — Now a piece of the melted wire drops to the bottom: I fear it is extinguished; but no, it burns again as bright as ever.

MRS. B.

It will burn till the wire is entirely consumed, provided the oxygen is not first expended: for you know it can burn only while there is oxygen to combine with it.

CAROLINE.

I never saw a more beautiful light. My eyes can hardly bear it! How astonishing to think that all this caloric was contained in the small quantity of gas and iron that was enclosed in the receiver; and that, without producing any sensible heat!

CAROLINE.

How wonderfully quick combustion goes on in pure oxygen gas! But pray, are these drops of burnt iron as heavy as the wire was before?

MRS. B.

They are even heavier; for the iron, in burning, has acquired exactly the weight of the oxygen which has disappeared, and is now combined with it. It has become an oxyd of iron.

CAROLINE.

I do not know what you mean by saying that the oxygen has *disappeared*, Mrs. B., for it was always invisible.

MRS. B.

True, my dear; the expression was incorrect. But though you could not see the oxygen gas, I believe you had no doubt of its presence, as the effect it produced on the wire was sufficiently evident.

CAROLINE.

Yes, indeed; yet you know it was the caloric, and not the oxygen gas itself, that dazzled us so much.

MRS. B.

You are not quite correct in your turn, in saying the caloric dazzled you; for caloric is invisible; it affects only the sense of feeling; it was the light which dazzled you.

CAROLINE.

True; but light and caloric are such constant companions, that it is difficult to separate them, even in idea.

MRS. B.

The easier it is to confound them, the more careful you should be in making the distinction.

CAROLINE.

But why has the water now risen, and filled part of the receiver?

MRS. B.

Indeed, Caroline, I did not suppose you would have asked such a question! I dare say, Emily, you can answer it.

EMILY.

Let me reflect . . . . . . The oxygen has combined with the wire; the caloric has escaped; consequently nothing can remain in the receiver, and the water will rise to fill the vacuum.

CAROLINE.

I wonder that I did not think of that. I wish that we had weighed the wire and the oxygen gas before combustion; we might then have found whether the weight of the oxyd was equal to that of both.

MRS. B.

You might try the experiment if you particularly wished it; but I can assure you, that, if accurately performed, it never fails to show that the additional weight of the oxyd is precisely equal to that

Fig. 1.

Fig. 2

d

c

b

Fig. 3.

Fig. 4.

Fig. 5.

Fig. 6.

Fig. 7.

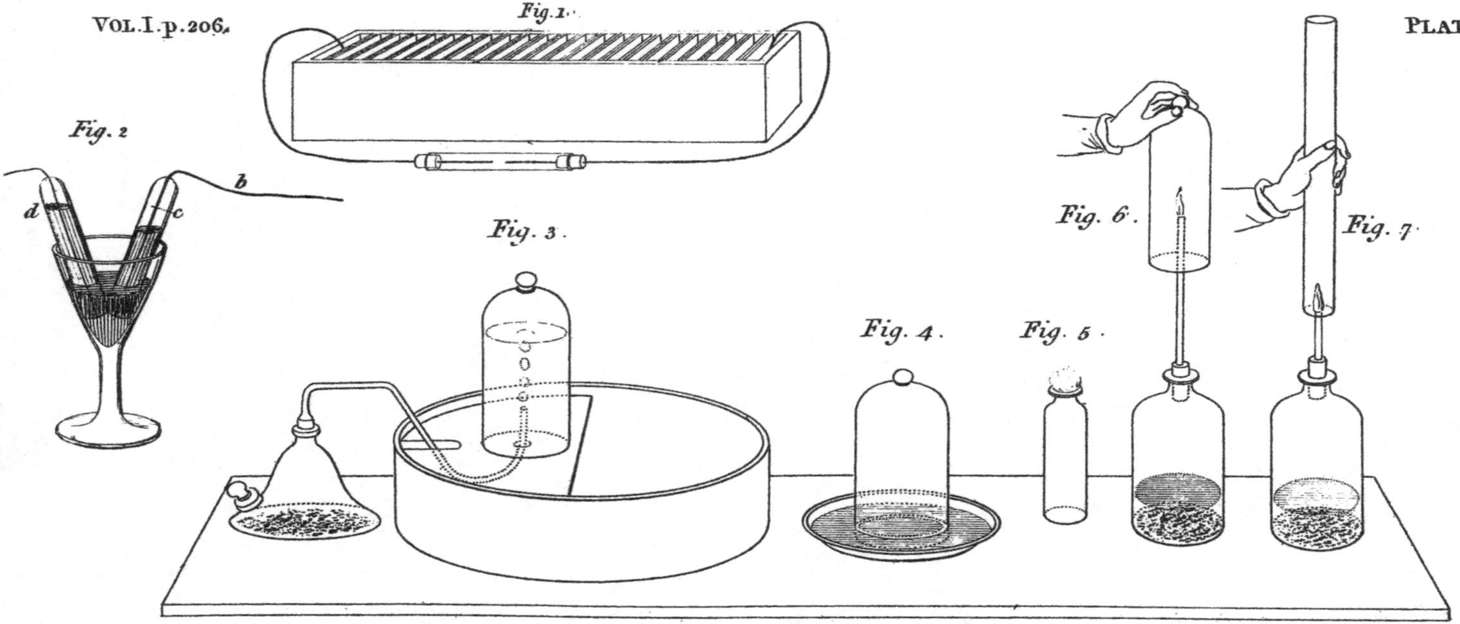

Fig. 1. Apparatus for the decomposition of water by the Voltaic Battery.

Fig. 3. Apparatus for preparing & collecting hydrogen gas.—Fig. 4. Receiver full of hydrogen gas inverted over water.—Fig. 5: Slow combustion of hydrogen gas.—Fig. 6: Apparatus for illustrating the formation of water by the combustion of hydrogen gas.—Fig. 7: Apparatus for producing harmonic sounds by the combustion of hydrogen gas.

Fig. 2. Apparatus for decomposing water by Voltaic Electricity & obtaining the gasses separate.

Drawn by the Author.

Engraved by Lowry

Published by Longman & C.º Oct.r 2nd 1809.

of the oxygen absorbed, whether the process has been a real combustion, or a simple oxygenation.

CAROLINE.

But this cannot be the case with combustions in general; for when any substance is burnt in the common air, so far from increasing in weight, it is evidently diminished, and sometimes entirely consumed.

MRS. B.

But what do you mean by the expression *consumed?* You cannot suppose that the smallest particle of any substance in nature can be actually destroyed. A compound body is decomposed by combustion; some of its constituent parts fly off in a gaseous form, while others remain in a concrete state; the former are called the *volatile*, the latter the *fixed products* of combustion. But if we collect the whole of them, we shall always find that they exceed the weight of the combustible body, by that of the oxygen which has combined with them during combustion.

EMILY.

In the combustion of a coal fire, then, I suppose that the ashes are what would be called the fixed product, and the smoke the volatile product?

MRS. B.

Yet when the fire burns best, and the quantity of volatile products should be the greatest, there is no smoke; how can you account for that?

EMILY.

Indeed I cannot; therefore I suppose that I was not right in my conjecture.

MRS. B.

Not quite: ashes, as you supposed, are a fixed product of combustion; but smoke, properly speaking, is not one of the volatile products, as it consists of some minute undecomposed particles of the coals that are carried off by the heated air without being burnt, and are either deposited in the form of soot, or dispersed by the wind. Smoke, therefore, ultimately, becomes one of the *fixed* products of combustion. And you may easily conceive that the stronger the fire is, the less smoke is produced, because the fewer particles escape combustion. On this principle depends the invention of Argand's Patent Lamps; a current of air is made to pass through the cylindrical wick of the lamp, by which means it is so plentifully supplied with oxygen, that scarcely a particle of oil escapes combustion, nor is there any smoke produced.

EMILY.

But what then are the volatile products of combustion?

MRS. B.

Various new compounds, with which you are not yet acquainted, and which being converted by caloric either into vapour or gas, are invisible; but they can be collected, and we shall examine them at some future period.

CAROLINE.

There are then other gases, besides the oxygen and nitrogen gases.

MRS. B.

Yes, several: any substance that can assume and maintain the form of an elastic fluid at the temperature of the atmosphere, is called a gas. We shall examine the several gases in their respective places; but we must now confine our attention to those that compose the atmosphere.

I shall show you another method of decomposing the atmosphere, which is very simple. In breathing, we retain a portion of the oxygen, and expire the nitrogen gas; so that if we breathe in a closed vessel, for a certain length of time, the air within it will be deprived of its oxygen gas. Which of you will make the experiment?

CAROLINE.

I should be very glad to try it.

MRS. B.

Very well; breathe several times through this glass tube into the receiver with which it is connected, until you feel that your breath is exhausted.

CAROLINE.

I am quite out of breath already!

MRS. B.

Now let us try the gas with a lighted taper.

EMILY.

It is very pure nitrogen gas, for the taper is immediately extinguished.

MRS. B.

That is not a proof of its being pure, but only of the absence of oxygen, as it is that principle alone which can produce combustion, every other gas being absolutely incapable of it.

EMILY.

In the methods which you have shown us, for decomposing the atmosphere, the oxygen always abandons the nitrogen; but is there no way of taking the nitrogen from the oxygen, so as to obtain the latter pure from the atmosphere?

MRS. B.

You must observe, that whenever oxygen is

taken from the atmosphere, it is by decomposing the oxygen gas; we cannot do the same with the nitrogen gas, because nitrogen has a stronger affinity for caloric than for any other known principle: it appears impossible therefore to separate it from the atmosphere by the power of affinities. But if we cannot obtain the oxygen gas, by this means, in its separate state, we have no difficulty (as you have seen) to procure it in its gaseous form, by taking it from those substances that have absorbed it from the atmosphere, as we did with the oxyd of manganese.

EMILY.

Can atmospherical air be recomposed, by mixing due proportions of oxygen and nitrogen gases?

MRS. B.

Yes: if about one part of oxygen gas be mixed with about four parts of nitrogen gas, atmospherical air is produced. *

EMILY.

The air, then, must be an oxyd of nitrogen?

MRS. B.

No, my dear; for there must be a chemical

* The proportion of oxygen in the atmosphere varies from 21 to 22 per cent.

combination between oxygen and nitrogen in order to produce an oxyd; whilst in the atmosphere these two substances are separately combined with caloric, forming two distinct gases, which are simply mixed in the formation of the atmosphere.

I shall say nothing more of oxygen and nitrogen at present, as we shall continually have occasion to refer to them in our future conversations. They are both very abundant in nature; nitrogen is the most plentiful in the atmosphere, and exists also in all animal substances; oxygen forms a constituent part, both of the animal and vegetable kingdoms, from which it may be obtained by a variety of chemical means. But it is now time to conclude our lesson. I am afraid you have learnt more to-day than you will be able to remember.

CAROLINE.

I assure you that I have been too much interested in it, ever to forget it. In regard to nitrogen there seems to be but little to remember; it makes a very insignificant figure in comparison to oxygen, although it composes a much larger portion of the atmosphere.

MRS. B.

Perhaps this insignificance you complain of may arise from the compound nature of nitrogen, for though I have hitherto considered it as a simple body, because it is not known in any natural process to be decomposed, yet from some experiments of Sir H. Davy, there appears to be reason for suspecting that nitrogen is a compound body, as shall see afterwards. But even in its simple state, it will not appear so insignificant when you are better acquainted with it; for though it seems to perform but a passive part in the atmosphere, and has no very striking properties, when considered in its separate state, yet you will see by-and-bye what a very important agent it becomes, when combined with other bodies. But no more of this at present; we must reserve it for its proper place.

CONVERSATION VII.

ON HYDROGEN.

CAROLINE.

THE next simple bodies we come to are CHLORINE and IODINE. Pray what kinds of substances are these; are they also invisible?

MRS. B.

No; for chlorine, in the state of gas, has a distinct greenish colour, and is therefore visible; and iodine, in the same state, has a beautiful claret-red colour. The knowledge of these two bodies, however, and the explanation of their properties, imply various considerations, which you would not yet be able to understand; we shall therefore defer their examination to some future conversation, and we shall pass on to the next simple substance, HYDROGEN, which we cannot, any more than oxygen, obtain in a visible or palpable form. We are acquainted with it only in its gaseous state, as we are with oxygen and nitrogen.

CAROLINE.

But in its gaseous state it cannot be called a

simple substance, since it is combined with heat and electricity?

MRS. B.

True, my dear; but as we do not know in nature of any substance which is not more or less combined with caloric and electricity, we are apt to say that a substance is in its pure state when combined with those agents only.

Hydrogen was formerly called *inflammable air*, as it is extremely combustible, and burns with a great flame. Since the invention of the new nomenclature, it has obtained the name of hydrogen, which is derived from two Greek words, the meaning of which is, *to produce water*.

EMILY.

And how does hydrogen produce water?

MRS. B.

By its combustion. Water is composed of eighty-five parts, by weight, of oxygen, combined with fifteen parts of hydrogen; or of two parts, by bulk of hydrogen gas, to one part of oxygen gas.

CAROLINE.

Really! is it possible that water should be a combination of two gases, and that one of these should be inflammable air! Hydrogen must be a most extraordinary gas that will produce both fire and water.

EMILY.

But I thought you said that combustion could take place in no gas but oxygen?

MRS. B.

Do you recollect what the process of combustion consists in?

EMILY.

In the combination of a body with oxygen, with disengagement of light and heat.

MRS. B.

Therefore when I say that hydrogen is combustible, I mean that it has an affinity for oxygen; but, like all other combustible substances, it cannot burn unless supplied with oxygen, and also heated to a proper temperature.

CAROLINE.

The simply mixing fifteen parts of hydrogen, with eighty-five parts of oxygen gas, will not, therefore, produce water?

MRS. B.

No; water being a much denser fluid than gases, in order to reduce these gases to a liquid, it is necessary to diminish the quantity of caloric or electricity which maintains them in an elastic form.

EMILY.

That I should think might be done by combining the oxygen and hydrogen together; for in combining they would give out their respective electricities in the form of caloric, and by this means would be condensed.

CAROLINE.

But you forget, Emily, that in order to make the oxygen and hydrogen combine, you must begin by elevating their temperature, which increases, instead of diminishing, their electric energies.

MRS. B.

Emily is, however, right; for though it is necessary to raise their temperature, in order to make them combine, as that combination affords them the means of parting with their electricities, it is eventually the cause of the diminution of electric energy.

CAROLINE.

You love to deal in paradoxes to-day, Mrs. B. — Fire, then, produces water?

MRS. B.

The combustion of hydrogen gas certainly does;

but you do not seem to have remembered the theory of combustion so well as you thought you would. Can you tell me what happens in the combustion of hydrogen gas?

CAROLINE.

The hydrogen combines with the oxygen, and their opposite electricities are disengaged in the form of caloric.—Yes, I think I understand it now—by the loss of this caloric, the gases are condensed into a liquid.

EMILY.

Water, then, I suppose, when it evaporates and incorporates with the atmosphere, is decomposed and converted into hydrogen and oxygen gases?

MRS. B.

No, my dear—there you are quite mistaken: the decomposition of water is totally different from its evaporation; for in the latter case (as you should recollect) water is only in a state of very minute division; and is merely suspended in the atmosphere, without any chemical combination, and without any separation of its constituent parts. As long as these remain combined, they form WATER, whether in a state of liquidity, or in that of an elastic fluid, as vapour, or under the solid form of ice.

In our experiments on latent heat, you may recollect that we caused water successively to pass through these three forms, merely by an increase or diminution of caloric, without employing any power of attraction, or effecting any decomposition.

CAROLINE.

But are there no means of decomposing water?

MRS. B.

Yes, several: charcoal, and metals, when heated red hot, will attract the oxygen from water, in the same manner as they will from the atmosphere.

CAROLINE.

Hydrogen, I see, is like nitrogen, a poor dependant friend of oxygen, which is continually forsaken for greater favourites.

MRS. B.

The connection, or friendship, as you choose to call it, is much more intimate between oxygen and hydrogen, in the state of water, than between oxygen and nitrogen, in the atmosphere; for, in the first case, there is a chemical union and condensation of the two substances; in the latter, they are simply mixed together in their gaseous state. You will find, however, that, in some cases, nitrogen is quite as intimately connected with oxygen, as hydrogen is.— But this is foreign to our present subject.

EMILY.

Water, then, is an oxyd, though the atmospherical air is not?

MRS. B.

It is not commonly called an oxyd, though, according to our definition, it may, no doubt, be referred to that class of bodies.

CAROLINE.

I should like extremely to see water decomposed.

MRS. B.

I can gratify your curiosity by a much more easy process than the oxydation of charcoal or metals: the decomposition of water by these latter means takes up a great deal of time, and is attended with much trouble; for it is necessary that the charcoal or metal should be made red hot in a furnace, that the water should pass over them in a state of vapour, that the gas formed should be collected over the water-bath, &c. In short, it is a very complicated affair. But the same effect may be produced with the greatest facility, by the action of the Voltaic battery, which this will give me an opportunity of exhibiting.

CAROLINE.

I am very glad of that, for I longed to see the power of this apparatus in decomposing bodies.

MRS. B.

For this purpose I fill this piece of glass-tube PLATE VIII. fig. 1.) with water, and cork it up at both ends; through one of the corks I introduce that wire of the battery which conveys the positive electricity; and the wire which conveys the negative electricity is made to pass through the other cork, so that the two wires approach each other sufficiently near to give out their respective electricities.

CAROLINE.

It does not appear to me that you approach the wires so near as you did when you made the battery act by itself.

MRS. B.

Water being a better conductor of electricity than air, the two wires will act on each other at a greater distance in the former than in the latter.

EMILY.

Now the electrical effect appears: I see small bubbles of air emitted from each wire.

MRS. B.

Each wire decomposes the water, the positive by combining with its oxygen which is negative, the negative by combining with its hydrogen which is positive.

CAROLINE.

That is wonderfully curious! But what are the small bubbles of air?

MRS. B.

Those that appear to proceed from the positive wire, are the result of the decomposition of the water by that wire. That is to say, the positive electricity having combined with some of the oxygen of the water, the particles of hydrogen which were combined with that portion of oxygen are set at liberty, and appear in the form of small bubbles of gas or air.

EMILY.

And I suppose the negative fluid having in the same manner combined with some of the hydrogen of the water, the particles of oxygen that were combined with it, are set free, and emitted in a gaseous form.

MRS. B.

Precisely so. But I should not forget to observe, that the wires used in this experiment are made of platina, a metal which is not capable of combining with oxygen; for otherwise the wire would combine with the oxygen, and the hydrogen alone would be disengaged.

CAROLINE.

But could not water be decomposed without the electric circle being completed? If, for instance, you immersed only the positive wire in the water, would it not combine with the oxygen, and the hydrogen gas be given out?

MRS. B.

No; for as you may recollect, the battery cannot act unless the circle be completed; since the positive wire will not give out its electricity, unless attracted by that of the negative wire.

CAROLINE.

I understand it now. — But look, Mrs. B., the decomposition of the water which has now been going on for some time, does not sensibly diminish its quantity — what is the reason of that?

MRS. B.

Because the quantity decomposed is so extremely small. If you compare the density of water with that of the gases into which it is resolved, you must be aware that a single drop of water is sufficient to produce thousands of such small bubbles as those you now perceive.

CAROLINE.

But in this experiment, we obtain the oxygen

and hydrogen gases mixed together. Is there any means of procuring the two gases separately?

MRS. B.

They can be collected separately with great ease, by modifying a little the experiment. Thus if instead of one tube, we employ two, as you see here, (c, d, PLATE VIII. fig. 2,) both tubes being closed at one end, and open at the other; and if after filling these tubes with water, we place them standing in a glass of water (e), with their open end downwards, you will see that the moment we connect the wires (a, b) which proceed upwards from the interior of each tube, the one with one end of the battery, and the other with the other end, the water in the tubes will be decomposed; hydrogen will be given out round the wire in the tube connected with the positive end of the battery, and oxygen in the other; and these gases will be evolved exactly in the proportions which I have before mentioned, namely, two measures of hydrogen for one of oxygen. We shall now begin the experiment, but it will be some time before any sensible quantity of the gases can be collected.

EMILY.

The decomposition of water in this way, slow as it is, is certainly very striking; but I confess that I should be still more gratified, if you could shew it us on a larger scale, and by a quicker process.

I am sorry that the decomposition of water by charcoal or metals is attended with so much inconvenience.

MRS. B.

Water may be decomposed by means of metals without any difficulty; but for this purpose the intervention of an acid is required. Thus, if we add some sulphuric acid (a substance with the nature of which you are not yet acquainted) to the water which the metal is to decompose, the acid disposes the metal to combine with the oxygen of the water so readily and abundantly, that no heat is required to hasten the process. Of this I am going to shew you an instance. I put into this bottle the water that is to be decomposed, as also the metal that is to effect that decomposition by combining with the oxygen, and the acid which is to facilitate the combination of the metal and the oxygen. You will see with what violence these will act on each other.

CAROLINE.

But what metal is it that you employ for this purpose?

MRS. B.

It is iron; and it is used in the state of filings, as these present a greater surface to the acid than a solid piece of metal. For as it is the surface of the metal which is acted upon by the acid, and is disposed to receive the oxygen produced by the decomposition of the water, it necessarily follows that the greater is the surface, the more considerable is the effect. The bubbles which are now rising are hydrogen gas ——

CAROLINE.

How disagreeably it smells!

MRS. B.

It is indeed unpleasant, though, I believe, not particularly hurtful. We shall not, however, suffer any more to escape, as it will be wanted for experiments. I shall, therefore, collect it in a glass-receiver, by making it pass through this bent tube, which will conduct it into the water-bath. (PLATE VIII. fig. 3.)

EMILY.

How very rapidly the gas escapes! it is perfectly transparent, and without any colour whatever. — Now the receiver is full ——

MRS. B.

We shall, therefore, remove it, and substitute another in its place. But you must observe, that when the receiver is full, it is necessary to keep it inverted with the mouth under water, otherwise the gas would escape. And in order that it may not be in the way, I introduce within the bath, under the water, a saucer, into which I slide the receiver, so that it can be taken out of the bath and conveyed any where, the water in the saucer being equally effectual in preventing its escape as that in the bath. (PLATE VIII. fig. 4.)

EMILY.

I am quite surprised to see what a large quantity of hydrogen gas can be produced by such a small quantity of water, especially as oxygen is the principal constituent of water.

MRS. B.

In weight it is; but not in volume. For though the proportion, by weight, is nearly six parts of oxygen to one of hydrogen, yet the proportion of the volume of the gases, is about one part of oxygen to two of hydrogen; so much heavier is the former than the latter.

CAROLINE.

But why is the vessel in which the water is decomposed so hot? As the water changes from a liquid to a gaseous form, cold should be produced instead of heat.

MRS. B.

No; for if one of the constituents of water is converted into a gas, the other becomes solid in combining with the metal.

EMILY.

In this case, then, neither heat nor cold should be produced?

MRS. B.

True; but observe that the sensible heat which is disengaged in this operation, is not owing to the decomposition of the water, but to an extrication of heat produced by the mixture of water and sulphuric acid. I will mix some water and sulphuric acid together in this glass, that you may feel the surprising quantity of heat that is disengaged by their union — now take hold of the glass——

CAROLINE.

Indeed I cannot; it feels as hot as boiling water. I should have imagined there would have been heat enough disengaged to have rendered the liquid solid.

MRS. B.

As, however, it does not produce that effect, we cannot refer this heat to the modification called latent heat. We may, however, I think, consider it as heat of capacity, as the liquid is condensed by its loss; and if you were to repeat the experiment, in a graduated tube, you would find that the two liquids, when mixed, occupy considerably less space than they did separately. — But we will reserve this to another opportunity, and attend at present to the hydrogen gas which we have been producing.

If I now set the hydrogen gas, which is contained in this receiver, at liberty all at once, and

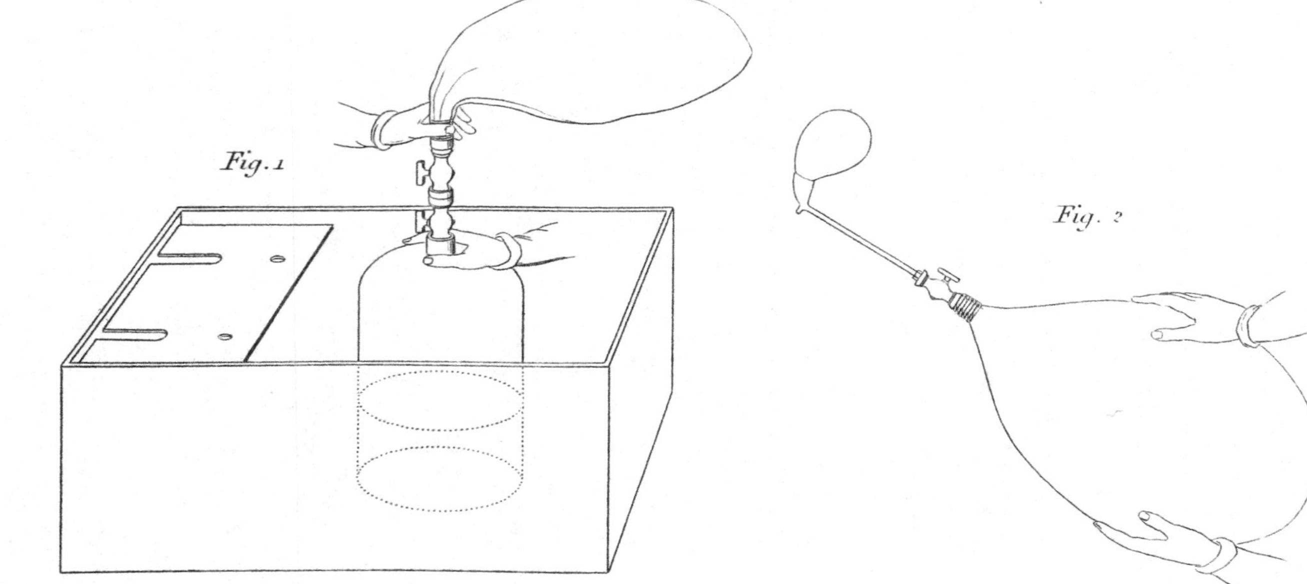

Fig. 1. Apparatus for transferring gases from a Receiver into a bladder.—Fig. 2. Apparatus for blowing Soap bubbles.

Drawn by the Author. *Published by Longman & Co. Octr. 2nd 1809.* *Engraved by Lowry.*

kindle it as soon as it comes in contact with the atmosphere, by presenting it to a candle, it will so suddenly and rapidly decompose the oxygen gas, by combining with its basis, that an explosion, or a *detonation* (as chemists commonly call it), will be produced. For this purpose, I need only take up the receiver, and quickly present its open mouth to the candle —— so . . . .

CAROLINE.

It produced only a sort of hissing noise, with a vivid flash of light. I had expected a much greater report.

MRS. B.

And so it would have been, had the gases been closely confined at the moment they were made to explode. If, for instance, we were to put in this bottle a mixture of hydrogen gas and atmospheric air; and if, after corking the bottle, we should kindle the mixture by a very small orifice, from the sudden dilatation of the gases at the moment of their combination, the bottle must either fly to pieces, or the cork be blown out with considerable violence.

CAROLINE.

But in the experiment which we have just seen, if you did not kindle the hydrogen gas, would it not equally combine with the oxygen?

MRS. B.

Certainly not; for, as I have just explained to you, it is necessary that the oxygen and hydrogen gases be burnt together, in order to combine chemically and produce water.

CAROLINE.

That is true; but I thought this was a different combination, for I see no water produced.

MRS. B.

The water resulting from this detonation was so small in quantity, and in such a state of minute division, as to be invisible. But water certainly was produced; for oxygen is incapable of combining with hydrogen in any other proportions than those that form water; therefore water must always be the result of their combination.

If, instead of bringing the hydrogen gas into sudden contact with the atmosphere (as we did just now) so as to make the whole of it explode the moment it is kindled, we allow but a very small surface of gas to burn in contact with the atmosphere, the combustion goes on quietly and gradually at the point of contact, without any detonation, because the surfaces brought together are too small for the immediate union of gases. The experiment is a very easy one. This phial, with a narrow neck, (Plate VIII. fig. 5.) is full of hydrogen gas, and is carefully corked. If I take out the cork without moving the phial, and quickly approach the candle to the orifice, you will see how different the result will be ——

EMILY.

How prettily it burns, with a blue flame! The flame is gradually sinking within the phial — now it has entirely disappeared. But does not this combustion likewise produce water?

MRS. B.

Undoubtedly. In order to make the formation of the water sensible to you, I shall procure a fresh supply of hydrogen gas, by putting into this bottle (PLATE VIII. fig. 6.) iron filings, water, and sulphuric acid, materials similar to those which we have just used for the same purpose. I shall then cork up the bottle, leaving only a small orifice in the cork, with a piece of glass-tube fixed to it, through which the gas will issue in a continued rapid stream.

CAROLINE.

I hear already the hissing of the gas through the tube, and I can feel a strong current against my hand.

MRS. B.

This current I am going to kindle with the candle — see how vividly it burns ——

EMILY.

It burns like a candle with a long flame. But why does this combustion last so much longer than in the former experiment?

MRS. B.

The combustion goes on uninterruptedly as long as the new gas continues to be produced. Now if I invert this receiver over the flame, you will soon perceive its internal surface covered with a very fine dew, which is pure water ——

CAROLINE.

Yes, indeed; the glass is now quite dim with moisture! How glad I am that we can *see* the water produced by this combustion.

EMILY.

It is exactly what I was anxious to see; for I confess I was a little incredulous.

MRS. B.

If I had not held the glass-bell over the flame, the water would have escaped in the state of vapour, as it did in the former experiment. We have here, of course, obtained but a very small quantity of water; but the difficulty of procuring a proper apparatus, with sufficient quantities of gases, prevents my showing it you on a larger scale.

The composition of water was discovered about the same period, both by Mr. Cavendish, in this country, and by the celebrated French chemist Lavoisier. The latter invented a very perfect and ingenious apparatus to perform, with great accuracy, and upon a large scale, the formation of water by the combination of oxygen and hydrogen gases. Two tubes, conveying due proportions, the one of oxygen, the other of hydrogen gas, are inserted at opposite sides of a large globe of glass, previously exhausted of air; the two streams of gas are kindled within the globe, by the electrical spark, at the point where they come in contact; they burn together, that is to say, the hydrogen combines with the oxygen, the caloric is set at liberty, and a quantity of water is produced exactly equal, in weight, to that of the two gases introduced into the globe.

CAROLINE.

And what was the greatest quantity of water ever formed in this apparatus?

MRS. B.

Several ounces; indeed, very nearly a pound, if I recollect right; but the operation lasted many days.

EMILY.

This experiment must have convinced all the world of the truth of the discovery. Pray, if improper proportions of the gases were mixed and set fire to, what would be the result?

MRS. B.

Water would equally be formed, but there would be a residue of either one or other of the gases, because, as I have already told you, hydrogen and oxygen will combine only in the proportions requisite for the formation of water.

EMILY.

Look, Mrs. B., our experiment with the Voltaic battery (PLATE VIII. fig. 2.) has made great progress; a quantity of gas has been formed in each tube, but in one of them there is twice as much gas as in the other.

MRS. B.

Yes; because, as I said before, water is composed of two volumes of hydrogen to one of oxygen — and if we should now mix these gases together and set fire to them by an electrical spark, both gases would entirely disappear, and a small quantity of water would be formed.

There is another curious effect produced by the combustion of hydrogen gas, which I shall show you, though I must acquaint you first, that I cannot well explain the cause of it. For this purpose, I must put some materials into our apparatus, in order to obtain a stream of hydrogen gas, just as we have done before. The process is already going on, and the gas is rushing through the tube — I shall now kindle it with the taper ——

EMILY.

It burns exactly as it did before —— What is the curious effect which you were mentioning?

MRS. B.

Instead of the receiver, by means of which we have just seen the drops of water form, we shall invert over the flame this piece of tube, which is about two feet in length, and one inch in diameter (PLATE VIII. fig. 7.); but you must observe that it is open at both ends.

EMILY.

What a strange noise it makes! something like the Æolian harp, but not so sweet.

CAROLINE.

It is very singular, indeed; but I think rather too powerful to be pleasing. And is not this sound accounted for?

MRS. B.

That the percussion of glass, by a rapid stream of gas, should produce a sound, is not extraordinary: but the sound here is so peculiar, that no other gas has a similar effect. Perhaps it is owing to a brisk vibratory motion of the glass, occasioned by the successive formation and condensation of small drops of water on the sides of the glass tube, and the air rushing in to replace the vacuum formed.*

CAROLINE.

How very much this flame resembles the burning of a candle.

MRS. B.

The burning of a candle is produced by much the same means. A great deal of hydrogen is contained in candles, whether of tallow or wax. This hydrogen being converted into gas by the heat of the candle, combines with the oxygen of the atmosphere, and flame and water result from this combination. So that, in fact, the flame of a candle is owing to the combustion of hydrogen gas. An elevation of temperature, such as is produced by a lighted match or taper, is required to give the first impulse to the combustion; but af-

* This ingenious explanation was first suggested by Dr. Delarive.— See Journals of the Royal Institution, vol. i. p. 259.

terwards it goes on of itself, because the candle finds a supply of caloric in the successive quantities of heat which results from the union of the two electricities given out by the gases during their combustion. But there are other circumstances connected with the combustion of candles and lamps, which I cannot explain to you till you are acquainted with *carbon*, which is one of their constituent parts. In general, however, whenever you see flame, you may infer that it is owing to the formation and burning of hydrogen gas *; for flame is the peculiar mode of burning hydrogen gas, which, with only one or two apparent exceptions, does not belong to any other combustible.

EMILY.

You astonish me! I understood that flame was the caloric produced by the union of the two electricities, in all combustions whatever?

MRS. B.

Your error proceeded from your vague and incorrect idea of flame; you have confounded it with light and caloric in general. Flame always implies caloric, since it is produced by the combustion of hydrogen gas; but all caloric does not

* Or rather, *hydro-carbonat*, a gas composed of hydrogen and carbon, which will be noticed under the head *Carbon*.

imply flame. Many bodies burn with intense heat without producing flame. Coals, for instance, burn with flame until all the hydrogen which they contain is evaporated; but when they afterwards become red hot, much more caloric is disengaged than when they produce flame.

CAROLINE.

But the iron wire, which you burnt in oxygen gas, appeared to me to emit flame; yet, as it was a simple metal, it could contain no hydrogen?

MRS. B.

It produced a sparkling dazzling blaze of light, but no real flame.

EMILY.

And what is the cause of the regular shape of the flame of a candle?

MRS. B.

The regular stream of hydrogen gas which exhales from its combustible matter.

CAROLINE.

But the hydrogen gas must, from its great levity, ascend into the upper regions of the atmosphere; why therefore does not the flame continue to accompany it?

MRS. B.

The combustion of the hydrogen gas is completed at the point where the flame terminates; it then ceases to be hydrogen gas, as it is converted by its combination with oxygen into watery vapour; but in a state of such minute division as to be invisible.

CAROLINE.

I do not understand what is the use of the wick of a candle, since the hydrogen gas burns so well without it?

MRS. B.

The combustible matter of the candle must be decomposed in order to emit the hydrogen gas, and the wick is instrumental in effecting this decomposition. Its combustion first melts the combustible matter, and . . . .

CAROLINE.

But in lamps the combustible matter is already fluid, and yet they also require wicks?

MRS. B.

I am going to add that, afterwards, the burning wick (by the power of capillary attraction) gradually draws up the fluid to the point where combustion takes place; for you must have observed that the wick does not burn quite to the bottom.

CAROLINE.

Yes; but I do not understand why it does not.

MRS. B.

Because the air has not so free an access to that part of the wick which is immediately in contact with the candle, as to the part just above, so that the heat there is not sufficient to produce its decomposition; the combustion therefore begins a little above this point.

CAROLINE.

But, Mrs. B., in those beautiful lights, called *gas-lights*, which are now seen in many streets, and will, I hope, be soon adopted every where. I can perceive no wick at all. How are these lights managed?

MRS. B.

I am glad you have put me in mind of saying a few words on this very useful and interesting improvement. In this mode of lighting, the gas is conveyed to the extremity of a tube, where it is kindled, and burns as long as the supply continues. There is, therefore, no occasion for a wick, or any other fuel whatever.

EMILY.

But how is all this gas procured in such large quantities?

MRS. B.

It is obtained from coal, by distillation. — Coal, when exposed to heat in a close vessel, is decomposed; and hydrogen, which is one of its constituents, rises in the state of gas, combined with another of its component parts, carbon, forming a compound gas, called *Hydrocarbonat*, the nature of which we shall again have an opportunity of noticing when we treat of carbon. This gas, like hydrogen, is perfectly transparent, invisible, and highly inflammable; and in burning it emits that vivid light which you have so often observed.

CAROLINE.

And does the process for procuring it require nothing but heating the coals, and conveying the gas through tubes?

MRS. B.

Nothing else; except that the gas must be made to pass, immediately at its formation, through two or three large vessels of water, in which it deposits some other ingredients, and especially water, tar, and oil, which also arise from the distillation of coals. The gas-light apparatus, therefore, consists simply in a large iron vessel, in which the coals are exposed to the heat of a furnace, — some reservoirs of water, in which the gas deposits its impurities, — and tubes that convey it to the desired spot, being propelled with uniform velocity through the tubes by means of a certain degree of pressure which is made upon the reservoir.

EMILY.

What an admirable contrivance! Do you not think, Mrs. B., that it will soon get into universal use?

MRS. B.

Most probably, as to the lighting of streets, offices, and public places, as it far surpasses any former invention for that purpose; but as to the interior of private houses, this mode of lighting has not yet been sufficiently tried to know whether it will be found generally desirable, either in regard to economy or convenience. It may, however, be considered as one of the happiest applications of chemistry to the comforts of life; and there is every reason to suppose that it will answer the full extent of public expectation.

I have another experiment to show you with hydrogen gas, which I think will entertain you. Have you ever blown bubbles with soap and water?

EMILY.

Yes, often, when I was a child; and I used to make them float in the air by blowing them upwards.

MRS. B.

We shall fill some such bubbles with hydrogen gas, instead of atmospheric air, and you will see with what ease and rapidity they will ascend, without the assistance of blowing, from the lightness of the gas. —Will you mix some soap and water whilst I fill this bladder with the gas contained in the receiver which stands on the shelf in the water-bath?

CAROLINE.

What is the use of the brass-stopper and turn-cock at the top of the receiver?

MRS. B.

It is to afford a passage to the gas when required. There is, you see, a similar stop-cock fastened to this bladder, which is made to fit that on the receiver. I screw them one on the other, and now turn the two cocks, to open a communication between the receiver and the bladder; then, by sliding the receiver off the shelf, and gently sinking it into the bath, the water rises in the receiver and forces the gas into the bladder. (Plate IX. fig. 1.)

CAROLINE.

Yes, I see the bladder swell as the water rises in the receiver.

MRS. B.

I think that we have already a sufficient quantity in the bladder for our purpose; we must be careful to stop both the cocks before we separate the bladder from the receiver, lest the gas should escape. — Now I must fix a pipe to the stopper of the bladder, and by dipping its mouth into the soap and water, take up a few drops — then I again turn the cock, and squeeze the bladder in order to force the gas into the soap and water at the mouth of the pipe. (PLATE IX. fig. 2.)

EMILY.

There is a bubble — but it bursts before it leaves the mouth of the pipe.

MRS. B.

We must have patience and try again; it is not so easy to blow bubbles by means of a bladder, as simply with the breath.

CAROLINE.

Perhaps there is not soap enough in the water; I should have had warm water, it would have dissolved the soap better.

EMILY.

Does not some of the gas escape between the bladder and the pipe?

MRS. B.

No, they are perfectly air tight; we shall succeed presently, I dare say.

CAROLINE.

Now a bubble ascends; it moves with the rapidity of a balloon. How beautifully it refracts the light!

EMILY.

It has burst against the ceiling — you succeed now wonderfully; but why do they all ascend and burst against the ceiling?

MRS. B.

Hydrogen gas is so much lighter than atmospherical air, that it ascends rapidly with its very light envelope, which is burst by the force with which it strikes the ceiling.

Air-balloons are filled with this gas, and if they carried no other weight than their covering, would ascend as rapidly as these bubbles.

CAROLINE.

Yet their covering must be much heavier than that of these bubbles?

MRS. B.

Not in proportion to the quantity of gas they contain. I do not know whether you have ever been present at the filling of a large balloon. The apparatus for that purpose is very simple. It consists of a number of vessels, either jars or barrels, in which the materials for the formation of the gas are mixed, each of these being furnished with a tube, and communicating with a long flexible pipe, which conveys the gas into the balloon.

EMILY.

But the fire-balloons which were first invented, and have been since abandoned, on account of their being so dangerous, were constructed, I suppose, on a different principle.

MRS. B.

They were filled simply with atmospherical air, considerably rarefied by heat; and the necessity of having a fire underneath the balloon, in order to preserve the rarefaction of the air within it, was the circumstance productive of so much danger.

If you are not yet tired of experiments, I have another to show you. It consists in filling soap-bubbles with a mixture of hydrogen and oxygen gases, in the proportions that form water; and afterwards setting fire to them.

EMILY.

They will detonate, I suppose?

MRS. B.

Yes, they will. As you have seen the method of transferring the gas from the receiver into the bladder, it is not necessary to repeat it. I have therefore provided a bladder which contains a due proportion of oxygen and hydrogen gases, and we have only to blow bubbles with it.

CAROLINE.

Here is a fine large bubble rising — shall I set fire to it with the candle?

MRS. B.

If you please . . . .

CAROLINE.

Heavens, what an explosion! — It was like the report of a gun: I confess it frightened me much. I never should have imagined it could be so loud.

EMILY.

And the flash was as vivid as lightning.

MRS. B.

The combination of the two gases takes place during that instant of time that you see the flash, and hear the detonation.

EMILY.

This has a strong resemblance to thunder and lightning.

MRS. B.

These phenomena, however, are generally of an electrical nature. Yet various meteorological effects may be attributed to accidental detonations of hydrogen gas in the atmosphere; for nature abounds with hydrogen: it constitutes a very considerable portion of the whole mass of water belonging to our globe, and from that source almost every other body obtains it. It enters into the composition of all animal substances, and of a great number of minerals; but it is most abundant in vegetables. From this immense variety of bodies, it is often spontaneously disengaged; its great levity makes it rise into the superior regions of the atmosphere; and when, either by an electrical spark, or any casual elevation of temperature, it takes fire, it may produce such meteors or luminous appearances as are occasionally seen in the atmosphere. Of this kind are probably those broad flashes which we often see on a summer-evening, without hearing any detonation.

EMILY.

Every flash, I suppose, must produce a quantity of water?

CAROLINE.

And this water, naturally, descends in the form of rain?

MRS. B.

That probably is often the case, though it is not a necessary consequence; for the water may be dissolved by the atmosphere, as it descends towards the lower regions, and remain there in the form of clouds.

The application of electrical attraction to chemical phenomena is likely to lead to many very interesting discoveries in meteorology; for electricity evidently acts a most important part in the atmosphere. This subject however, is, as yet, not sufficiently developed for me to venture enlarging upon it. The phenomena of the atmosphere are far from being well understood; and even with the little that is known, I am but imperfectly acquainted.

But before we take leave of hydrogen, I must not omit to mention to you a most interesting discovery of Sir H. Davy, which is connected with this subject.

CAROLINE.

You allude, I suppose, to the new miner's lamp, which has of late been so much talked of? I have long been desirous of knowing what that discovery was, and what purpose it was intended to answer.

MRS. B.

It often happens in coal-mines, that quantities of the gas, called by chemists *hydro-carbonat*, or by the miners *fire-damp*, (the same from which the gas-lights are obtained,) ooze out from fissures in the beds of coal, and fill the cavities in which the men are at work; and this gas being inflammable, the consequence is, that when the men approach those places with a lighted candle, the gas takes fire, and explosions happen which destroy the men and horses employed in that part of the colliery, sometimes in great numbers.

EMILY.

What tremendous accidents these must be! But whence does that gas originate?

MRS. B.

Being the chief product of the combustion of coal, no wonder that inflammable gas should occasionally appear in situations in which this mineral abounds, since there can be no doubt that processes of combustion are frequently taking place at a great depth under the surface of the earth; and therefore those accumulations of gas may arise either from combustions actually going on, or from former combustions, the gas having perhaps been confined there for ages.

CAROLINE.

And how does Sir H. Davy's lamp prevent those dreadful explosions?

MRS. B.

By a contrivance equally simple and ingenious; and one which does no less credit to the philosophical views from which it was deduced, than to the philanthropic motives from which the enquiry sprung. The principle of the lamp is shortly this: It was ascertained, two or three years ago, both by Mr. Tennant and by Sir Humphry himself, that the combustion of inflammable gas could not be propagated through small tubes; so that if a jet of an inflammable gaseous mixture, issuing from a bladder or any other vessel, through a small tube, be set fire to, it burns at the orifice of the tube, but the flame never penetrates into the vessel. It is upon this fact that Sir Humphry's safety-lamp is founded.

EMILY.

But why does not the flame ever penetrate through the tube into the vessel from which the gas issues, so as to explode at once the whole of the gas?

MRS. B.

Because, no doubt, the inflamed gas is so much cooled in its passage through a small tube as to cease to burn before the combustion reaches the reservoir.

CAROLINE.

And how can this principle be applied to the construction of a lamp?

MRS. B.

Nothing easier. You need only suppose a lamp enclosed all round in glass or horn, but having a number of small open tubes at the bottom, and others at the top, to let the air in and out. Now, if such a lamp or lanthorn be carried into an atmosphere capable of exploding, an explosion or combustion of the gas will take place *within* the lamp; and although the vent afforded by the tubes will save the lamp from bursting, yet, from the principle just explained, the combustion will not be propagated to the external air through the tubes, so that no farther consequence will ensue.

EMILY.

And is that all the mystery of that valuable lamp?

MRS. B.

No; in the early part of the enquiry a lamp of this kind was actually proposed; but it was but a rude sketch compared to its present state of improvement. Sir H. Davy, after a succession of trials, by which he brought his lamp nearer and nearer to perfection, at last conceived the happy idea that if the lamp were surrounded with a wire-work or wire-gauze, of a close texture, instead of glass or horn, the tubular contrivance I have just described would be entirely superseded, since each of the interstices of the gauze would act as a tube in preventing the propagation of explosions; so that this pervious metallic covering would answer the various purposes of transparency, of permeability to air, and of protection against explosion. This idea, Sir Humphry immediately submitted to the test of experiment, and the result has answered his most sanguine expectations, both in his laboratory and in the collieries, where it has already been extensively tried. And he has now the happiness of thinking that his invention will probably be the means of saving every year a number of lives, which would have been lost in digging out of the bowels of the earth one of the most valuable necessaries of life. Here is one of these lamps, every part of which you will at once comprehend. (See Plate X. fig. 1.)

CAROLINE.

How very simple and ingenious! But I do not yet well see why an explosion taking place within the lamp should not communicate to the external air around it, through the interstices of the wire?

MRS. B.

This has been and is still a subject of wonder, even to philosophers; and the only mode they have of explaining it is, that flame or ignition cannot pass through a fine wire-work, because the metallic wire cools the flame sufficiently to extinguish it in passing through the gauze. This property of the wire-gauze is quite similar to that of the tubes which I mentioned on introducing the subject; for you may consider each interstice of the gauze as an extremely short tube of a very small diameter.

EMILY.

But I should expect the wire would often become red-hot, by the burning of the gas within the lamp?

MRS. B.

And this is actually the case, for the top of the lamp is very apt to become red-hot. But, fortunately, inflammable gaseous mixtures cannot be exploded by red-hot wire, the intervention of actual flame being required for that purpose; so that the wire does not set fire to the explosive gas around it.

EMILY.

I can understand that; but if the wire be red-hot, how can it cool the flame within, and prevent its passing through the gauze?

MRS. B.

The gauze, though red-hot, is not so hot as the flame by which it has been heated; and as metallic wire is a good conductor, the heat does not much accumulate in it, as it passes off quickly to the other parts of the lamp, as well as to any contiguous bodies.

CAROLINE.

This is indeed a most interesting discovery, and one which shows at once the immense utility with which science may be practically applied to some of the most important purposes.

CONVERSATION VIII.

ON SULPHUR AND PHOSPHORUS.

MRS. B.

Sulphur is the next substance that comes under our consideration.. It·differs in one essential point from the preceding, as it exists in a solid form at the temperature of the atmosphere.

CAROLINE.

I am glad that we have at last a solid body to examine; one that we can see and touch. Pray, is it not with sulphur that the points of matches are covered, to make them easily kindle?

MRS. B.

Yes, it is; and you therefore already know that sulpur is a very combustible substance. It is seldom discovered in nature in a pure unmixed state; so great is its affinity for other substances, that it is almost constantly found combined with some of them. It is most commonly united with metals, under various forms, and is separated from them by a very simple process. It exists likewise in many mineral waters, and some vegetables yield it in various proportions, especially those of the cruciform tribe. It is also found in animal matter; in short, it may be discovered in greater or less quantity, in the mineral, vegetable, and animal kingdoms.

EMILY.

I have heard of *flowers of sulphur*, are they the produce of any plant?

MRS. B.

By no means: they consist of nothing more than common sulphur, reduced to a very fine powder by a process called *sublimation*.—You see some of it in this phial; it is exactly the same substance as this lump of sulphur, only its colour is a paler yellow, owing to its state of very minute division.

EMILY.

Pray what is sublimation?

MRS. B.

It is the evaporation, or, more properly speaking, the volatilisation of solid substances, which, in cooling, condense again in a concrete form.

The process, in this instance, must be performed in a closed vessel, both to prevent combustion, which would take place if the access of air were not carefully precluded, and likewise in order to collect the substance after the operation. As it is rather a slow process, we shall not try the experiment now; but you will understand it perfectly if I show you the apparatus used for the purpose. (PLATE XI. fig. 1.) Some lumps of sulphur are put into a receiver of this kind, which is called a *cucurbit*. Its shape, you see, somewhat resembles that of a pear, and is open at the top, so as to adapt itself exactly to a kind of conical receiver of this sort, called the head. The cucurbit, thus covered with its head, is placed over a sand-bath; this is nothing more than a vessel full of sand, which is kept heated by a furnace, such as you see here, so as to preserve the apparatus in a moderate and uniform temperature. The sulphur then soon begins to melt, and immediately after this, a thick white smoke rises, which is gradually deposited within the head, or upper part of the apparatus, where it condenses against the sides, somewhat in the form of a vegetation, whence it has obtained the name of flowers of sulphur. This apparatus, which is called an *alembic*, is highly useful in all kinds of distillations, as you will see when we come to treat of those operations. Alembics are not commonly made of glass, like this, which is applicable only to distillations upon a very small scale. Those used in manufactures are generally made of copper, and are, of course, considerably larger. The principal construction, however, is always the same, although their shape admits of some variation.

CAROLINE.

What is the use of that neck, or tube, which bends down from the upper piece of the apparatus?

MRS. B.

It is of no use in sublimations; but in distillations (the general object of which is to evaporate, by heat, in closed vessels, the volatile parts of a compound body, and to condense them again into a liquid,) it serves to carry off the condensed fluid, which otherwise would fall back into the cucurbit. But this is rather foreign to our present subject. Let us return to the sulphur. You now perfectly understand, I suppose, what is meant by sublimation?

EMILY.

I believe I do. Sublimation appears to consist in destroying, by means of heat, the attraction of aggregation of the particles of a solid body, which are thus volatilised; and as soon as they lose the caloric which produced that effect, they are deposited in the form of a fine powder.

CAROLINE.

It seems to me to be somewhat similar to the transformation of water into vapour, which returns to its liquid state when deprived of caloric.

EMILY.

There is this difference, however, that the sulphur does not return to its former state, since, instead of lumps, it changes to a fine powder.

MRS. B.

Chemically speaking, it is exactly the same substance, whether in the form of lump or powder. For if this powder be melted again by heat, it will, in cooling, be restored to the same solid state in which it was before its sublimation.

CAROLINE.

But if there be no real change, produced by the sublimation of the sulphur, what is the use of that operation?

MRS. B.

It divides the sulphur into very minute parts, and thus disposes it to enter more readily into combination with other bodies. It is used also as a means of purification.

CAROLINE.

Sublimation appears to me like the beginning of combustion, for the completion of which one circumstance only is wanting, the absorption of oxygen.

MRS. B.

But that circumstance is every thing. No essential alteration is produced in sulphur by sublimation; whilst in combustion it combines with the oxygen, and forms a new compound totally different in every respect from sulphur in its pure state. — We shall now *burn* some sulphur, and you will see how very different the result will be. For this purpose I put a small quantity of flowers of sulphur into this cup, and place it in a dish, into which I have poured a little water: I now set fire to the sulphur with the point of this hot wire; for its combustion will not begin unless its temperature be considerably raised. — You see that it burns with a faint blueish flame; and as I invert over it this receiver, white fumes arise from the sulphur, and fill the vessel. — You will soon perceive that the water is rising within the receiver, a little above its level in the plate. — Well, Emily, can you account for this?

EMILY.

I suppose that the sulphur has absorbed the oxygen from the atmospherical air within the receiver, and that we shall find some oxygenated sulphur in the cup. As for the white smoke, I am quite at a loss to guess what it may be.

MRS. B.

Your first conjecture is very right: but you are mistaken in the last; for nothing will be left in the cup. The white vapour is the oxygenated sulphur, which assumes the form of an elastic fluid of a pungent and offensive smell, and is a powerful acid. Here you see a chemical combination of oxygen and sulphur, producing a true gas, which would continue such under the pressure and at the temperature of the atmosphere, if it did not unite with the water in the plate, to which it imparts its acid taste, and all its acid properties. — You see, now, with what curious effects the combustion of sulphur is attended.

CAROLINE.

This is something quite new; and I confess that I do not perfectly understand why the sulphur turns acid.

MRS. B.

It is because it unites with oxygen, which is the acidifying principle. And, indeed, the word *oxygen* is derived from two Greek words signifying *to produce an acid.*

CAROLINE.

Why, then, is not water, which contains such a quantity of oxygen, acid?

MRS. B.

Because hydrogen, which is the other constituent of water, is not susceptible of acidification. — I believe it will be necessary, before we proceed further, to say a few words of the general nature of acids, though it is rather a deviation from our plan of examining the simple bodies separately, before we consider them in a state of combination.

Acids may be considered as a peculiar class of *burnt* bodies, which during their combustion, or combination with oxygen, have acquired very characteristic properties. They are chiefly discernible by their sour taste, and by turning red most of the blue vegetable colours. These two properties are common to the whole class of acids; but each of them is distinguished by other peculiar qualities. Every acid consists of some particular substance, (which constitutes its basis, and is different in each,) and of oxygen, which is common to them all.

EMILY.

But I do not clearly see the difference between acids and oxyds.

MRS. B.

Acids were, in fact, oxyds, which, by the addition of a sufficient quantity of oxygen, have been converted into acids. For acidification, you must observe, always implies previous oxydation, as a body must have combined with the quantity of oxygen requisite to constitute it an oxyd, before it can combine with the greater quantity that is necessary to render it an acid.

CAROLINE.

Are all oxyds capable of being converted into acids?

MRS. B.

Very far from it; it is only certain substances which will enter into that peculiar kind of union with oxygen that produces acids, and the number of these is proportionally very small; but all burnt bodies may be considered as belonging either to the class of oxyds, or to that of acids. At a future period, we shall enter more at large into this subject. At present, I have but one circumstance further to point out to your observation respecting acids: it is, that most of them are susceptible of two degrees of acidification, according to the different quantities of oxygen with which their basis combines.

EMILY.

And how are these two degrees of acidification distinguished?

MRS. B.

By the peculiar properties which result from them. The acid we have just made is the first or weakest degree of acidification, and is called *sulphureous acid*; if it were fully saturated with oxygen, it would be called *sulphuric acid.* You must therefore remember, that in this, as in all acids, the first degree of acidification is expressed by the termination in *ous;* the stronger, by the termination in *ic.*

CAROLINE.

And how is the sulphuric acid made?

MRS. B.

By burning sulphur in pure oxygen gas, and thus rendering its combustion much more complete. I have provided some oxygen gas for this purpose; it is in that bottle, but we must first decant the gas into the glass receiver which stands on the shelf in the bath, and is full of water.

CAROLINE.

Pray, let me try to do it, Mrs. B.

MRS. B.

It requires some little dexterity—hold the bottle completely under water, and do not turn the mouth upwards, till it is immediately under the aperture in the shelf, through which the gas is to pass into the receiver, and then turn it up gradually.—Very well, you have only let a few bubbles escape, and that must be expected at a first trial.—Now I shall put this piece of sulphur into the receiver, through the opening at the top, and introduce along with it a small piece of lighted tinder to set fire to it. — This requires being done very quickly, lest the atmospherical air should get in, and mix with the pure oxygen gas.

EMILY.

How beautifully it burns!

CAROLINE.

But it is already buried in the thick vapour. This, I suppose, is sulphuric acid?

EMILY.

Are these acids always in a gaseous state?

MRS. B.

Sulphureous acid, as we have already observed, is a permanent gas, and can be obtained in a liquid form only by condensing it in water. In its pure state, the sulphureous acid is invisible, and it now appears in the form of a white smoke, from its combining with the moisture. But the vapour of sulphuric acid, which you have just seen to rise during the combustion, is not a gas, but only a vapour, which condenses into liquid sulphuric acid, by losing its caloric. But it appears from Sir H. Davy's experiments, that this formation and condensation of sulphuric acid requires the presence of water, for which purpose the vapour is received into cold water, which may afterwards be separated from the acid by evaporation.

Sulphur has hitherto been considered as a simple substance; but Sir H. Davy has suspected that it contains a small portion of hydrogen, and perhaps also of oxygen.

On submitting sulphur to the action of the Voltaic battery, he observed that the negative wire gave out hydrogen; and the existence of hydrogen in sulphur was rendered still more probable by his observing that a small quantity of water was produced during the combustion of sulphur.

EMILY.

And pray of what nature is sulphur when perfectly pure?

MRS. B.

Sulphur has probably never been obtained perfectly free from combination, so that its radical may possibly possess properties very different from those of common sulphur. It has been suspected to be of a metallic nature; but this is mere conjecture.

Before we quit the subject of sulphur, I must tell you that it is susceptible of combining with a great variety of substances, and especially with hydrogen, with which you are already acquainted. Hydrogen gas can dissolve a small portion of it.

EMILY.

What! can a gas dissolve a solid substance?

MRS. B.

Yes; a solid substance may be so minutely divided by heat, as to become soluble in a gas: and there are several instances of it. But you must observe, that, in this case, a chemical union or combination of the sulphur with the hydrogen gas is produced. In order to effect this, the sulphur must be strongly heated in contact with the gas; the heat reduces the sulphur to such a state of extreme division, and diffuses it so thoroughly through the gas, that they combine and incorporate together. And as a proof that there must be a chemical union between the sulphur and the gas, it is sufficient to remark that they are not separated when the sulphur loses the caloric by which it was volatilized. Besides, it is evident, from the peculiar fetid smell of this gas, that it is a new compound totally different from either of its constituents; it is called *sulphuretted hydrogen gas*, and is contained in great abundance in sulphureous mineral waters.

CAROLINE.

Are not the Harrogate waters of this nature?

MRS. B.

Yes; they are naturally impregnated with sulphuretted hydrogen gas, and there are many other springs of the same kind, which shows that this gas must often be formed in the bowels of the earth by spontaneous processes of nature.

CAROLINE.

And could not such waters be made artificially by impregnating common water with this gas?

MRS. B.

Yes; they can be so well imitated, as perfectly to resemble the Harrogate waters.

Sulphur combines likewise with phosphorus, and with the alkalies, and alkaline earths, substances with which you are yet unacquainted. We cannot, therefore, enter into these combinations at present. In our next lesson we shall treat of phosphorus.

EMILY.

May we not begin that subject to-day; this lesson has been so short?

MRS. B.

I have no objection, if you are not tired. What do you say, Caroline?

CAROLINE.

I am as desirous as Emily of prolonging the lesson to-day, especially as we are to enter on a new subject; for I confess that sulphur has not appeared to me so interesting as the other simple bodies.

MRS. B.

Perhaps you may find phosphorus more entertaining You must not, however, be discouraged when you meet with some parts of a study less amusing than others; it would answer no good purpose to select the most pleasing parts, since, if we did not proceed with some method, in order to acquire a general idea of the whole, we could scarcely expect to take interest in any particular subjects.

PHOSPHORUS.

Phosphorus is considered as a simple body; though, like sulphur, it has been suspected of containing hydrogen. It was not known by the earlier chemists. It was first discovered by Brandt, a chemist of Hamburgh, whilst employed in researches after the philosopher's stone; but the method of obtaining it remained a secret till it was a second time discovered both by Kunckel and Boyle, in the year 1680. You see a specimen of phosphorus in this phial; it is generally moulded into small sticks of a yellowish colour, as you find it here.

CAROLINE.

I do not understand in what the discovery consisted; there may be a secret method of making an artificial composition, but how can you talk of *making* a substance which naturally exists?

MRS. B.

A body may exist in nature so closely combined with other substances, as to elude the observation of chemists, or render it extremely difficult to obtain it in its separate state. This is the case with phosphorus, which is always so intimately combined with other substances, that its existence remained unnoticed till Brandt discovered the means of obtaining it free from other combinations. It is found in all animal substances, and is now chiefly extracted from bones, by a chemical process. It exists also in some plants, that bear a strong analogy to animal matter in their chemical composition.

EMILY.

But is it never found in its pure separate state?

MRS. B.

Never, and this is the reason that it has remained so long undiscovered.

Phosphorus is eminently combustible; it melts and takes fire at the temperature of one hundred degrees, and absorbs in its combustion nearly once and a half its own weight of oxygen.

CAROLINE.

What! will a pound of phosphorus consume a pound and half of oxygen?

MRS. B.

So it appears from accurate experiments. I can show you with what violence it combines with oxygen, by burning some of it in that gas. We must manage the experiment in the same manner as we did the combustion of sulphur. You see I am obliged to cut this little bit of phosphorus under water, otherwise there would be danger of its taking fire by the heat of my fingers. I now put into the receiver, and kindle it by means of a hot wire.

EMILY.

What a blaze! I can hardly look at it. I never saw any thing so brilliant. Does it not hurt your eyes, Caroline?

CAROLINE.

Yes; but still I cannot help looking at it. A prodigious quantity of oxygen must indeed be absorbed, when so much light and caloric are disengaged!

MRS. B.

In the combustion of a pound of phosphorus, a sufficient quantity of caloric is set free to melt upwards of a hundred pounds of ice; this has been computed by direct experiments with the calorimeter.

EMILY.

And is the result of this combustion, like that of sulphur, an acid?

MRS. B.

Yes; phosphoric acid. And had we duly proportioned the phosphorus and the oxygen, they would have been completely converted into phosphoric acid, weighing together, in this new state, exactly the sum of their weights separately. The water would have ascended into the receiver, on account of the vacuum formed, and would have filled it entirely. In this case, as in the combustion of sulphur, the acid vapour formed is absorbed and condensed in the water of the receiver. But when this combustion is performed without any water or moisture being present, the acid then appears in the form of concrete whitish flakes, which are, however, extremely ready to melt upon the least admission of moisture.

EMILY.

Does phosphorus, in burning in atmospherical

air, produce, like sulphur, a weaker sort of the same acid?

MRS. B.

No: for it burns in atmospherical air, nearly at the same temperature as in pure oxygen gas; and it is in both cases so strongly disposed to combine with the oxygen, that the combustion is perfect, and the product similar; only in atmospherical air, being less rapidly supplied with oxygen, the process is performed in a slower manner.

CAROLINE.

But is there no method of acidifying phosphorus in a slighter manner, so as to form *phosphorus* acid?

MRS. B.

Yes, there is. When simply exposed to the atmosphere, phosphorus undergoes a kind of slow combustion at any temperature above zero.

EMILY.

But is not the process in this case rather an oxydation than a combustion? For if the oxygen is too slowly absorbed for a sensible quantity of light and heat to be disengaged, it is not a true combustion.

MRS. B.

The case is not as you suppose: a faint light is

emitted which is very discernible in the dark; but the heat evolved is not sufficiently strong to be sensible; a whitish vapour arises from this combustion, which, uniting with water, condenses into liquid phosphorus acid.

CAROLINE.

Is it not very singular that phosphorus should burn at so low a temperature in atmospherical air, whilst it does not burn in pure oxygen without the application of heat?

MRS. B.

So it at first appears. But this circumstance seems to be owing to the nitrogen gas of the atmosphere. This gas dissolves small particles of phosphorus, which being thus minutely divided and diffused in the atmospherical air, combines with the oxygen, and undergoes this slow combustion. But the same effect does not take place in oxygen gas, because it is not capable of dissolving phosphorus; it is therefore necessary, in this case, that heat should be applied to effect that division of particles, which, in the former instance, is produced by the nitrogen.

EMILY.

I have seen letters written with phosphorus, which are invisible by day-light, but may be read in the dark by their own light. They look as if they were written with fire; yet they do not seem to burn.

MRS. B.

But they do really burn; for it is by their slow combustion that the light is emitted; and phosphorus acid is the result of this combustion.

Phosphorus is sometimes used as a test to estimate the purity of atmospherical air. For this purpose, it is burnt in a graduated tube, called an *Eudiometer* (PLATE XI. fig. 2.), and from the quantity of air which the phosphorus absorbs, the proportion of oxygen in the air examined is deduced; for the phosphorus will absorb all the oxygen, and the nitrogen alone will remain.

EMILY.

And the more oxygen is contained in the atmosphere, the purer, I suppose, it is esteemed?

MRS. B.

Certainly. Phosphorus, when melted, combines with a great variety of substances. With sulphur it forms a compound so extremely combustible, that it immediately takes fire on coming in contact with the air. It is with this composition that phosphoric matches are prepared, which kindle as soon as they are taken out of their case and are exposed to the air.

EMILY.

I have a box of these curious matches; but I have observed, that in very cold weather, they will not take fire without being previously rubbed.

MRS. B.

By rubbing them you raise their temperature; for, you know, friction is one of the means of extricating heat.

EMILY.

Will phosphorus combine with hydrogen gas, as sulphur does?

MRS. B.

Yes; and the compound gas which results from this combination has a smell still more fetid than the sulphuretted hydrogen; it resembles that of garlic.

The *phosphoretted hydrogen gas* has this remarkable peculiarity, that it takes fire spontaneously in the atmosphere, at any temperature. It is thus, probably, that are produced those transient flames, or flashes of light, called by the vulgar *Will-of-the Whisp*, or more properly *Ignes-fatui*, which are often seen in church-yards, and places where the putrefactions of animal matter exhale phosphorus and hydrogen gas.

CAROLINE.

Country people, who are so much frightened by

those appearances, would soon be reconciled to them, if they knew from what a simple cause they proceed.

MRS. B.

There are other combinations of phosphorus that have also very singular properties, particularly that which results from its union with lime.

EMILY.

Is there any name to distinguish the combination of two substances, like phosphorus and lime, neither of which are oxygen, and which cannot therefore produce either an oxyd or an acid?

MRS. B.

The names of such combinations are composed from those of their ingredients, merely by a slight change in their termination. Thus the combination of sulphur with lime is called a *sulphuret*, and that of phosphorus, a *phosphuret of lime*. This latter compound, I was going to say, has the singular property of decomposing water, merely by being thrown into it. It effects this by absorbing the oxygen of water, in consequence of which bubbles of hydrogen gas ascend, holding in solution a small quantity of phosphorus.

EMILY.

These bubbles then are *phosphoretted hydrogen gas?*

MRS. B.

Yes; and they produce the singular appearance of a flash of fire issuing from water, as the bubbles kindle and detonate on the surface of the water, at the instant that they come in contact with the atmosphere.

CAROLINE.

Is not this effect nearly similar to that produced by the combination of phosphorus and sulphur, or, more properly speaking, the *phosphuret of sulphur?*

MRS. B.

Yes; but the phenomenon appears more extraordinary in this case, from the presence of water, and from the gaseous form of the combustible compound. Besides, the experiment surprises by its great simplicity. You only throw a piece of phosphoret of lime into a glass of water, and bubbles of fire will immediately issue from it.

CAROLINE.

Cannot we try the experiment?

MRS. B.

Very easily: but we must do it in the open air; for the smell of the phosphorated hydrogen gas is so extremely fetid, that it would be intolerable in the house. But before we leave the room, we may produce, by another process, some bubbles of the same gas, which are much less offensive.

There is in this little glass retort a solution of potash in water; I add to it a small piece of phosphorus. We must now heat the retort over the lamp, after having engaged its neck under water —you see it begins to boil; in a few minutes bubbles will appear, which take fire and detonate as they issue from the water.

CAROLINE.

There is one—and another. How curious it is! —But I do not understand how this is produced.

MRS. B.

It is the consequence of a display of affinities too complicated, I fear, to be made perfectly intelligible to you at present.

In a few words, the reciprocal action of the potash, phosphorus, caloric, and water are such, that some of the water is decomposed, and the hydrogen gas thereby formed carries off some minute particles of phosphorus, with which it forms phosphoretted hydrogen gas, a compound which spontaneously takes fire at almost any temperature.

EMILY.

What is that circular ring of smoke which slowly rises from each bubble after its detonation.

MRS. B.

It consists of water and phosphoric acid in vapour, which are produced by the combustion of hydrogen and phosphorus.

CONVERSATION IX.

ON CARBON.

CAROLINE.

To-day, Mrs. B., I believe we are to learn the nature and properties of carbon. This substance is quite new to me; I never heard it mentioned before.

MRS. B.

Not so new as you imagine; for carbon is nothing more than charcoal in a state of purity, that is to say, unmixed with any foreign ingredients.

CAROLINE.

But charcoal is made by art, Mrs. B., and a body consisting of one simple substance cannot be fabricated?

MRS. B.

You again confound the idea, of making a simple body, with that of separating it from a compound. The chemical processes by which a simple body is obtained in a state of purity, consist in *unmaking* the compound in which it is contained, in order to separate from it the simple substance in question. The method by which charcoal is usually obtained, is, indeed, commonly called *making* it; but, upon examination, you will find this process to consist simply in separating it from other substances with which it is found combined in nature.

Carbon forms a considerable part of the solid matter of all organised bodies; but it is most abundant in the vegetable creation, and it is chiefly obtained from wood. When the oil and water (which are other constituents of vegetable matter) are evaporated, the black, porous, brittle substance that remains, is charcoal.

CAROLINE.

But if heat be applied to the wood in order to evaporate the oil and water, will not the temperature of the charcoal be raised so as to make it burn; and if it combines with oxygen, can we any longer call it pure?

MRS. B.

I was going to say, that, in this operation, the air must be excluded.

CAROLINE.

How then can the vapour of the oil and water fly off?

MRS. B.

In order to produce charcoal in its purest state (which is, even then, but a less imperfect sort of carbon), the operation should be performed in an earthen retort. Heat being applied to the body of the retort, the evaporable part of the wood will escape through its neck, into which no air can penetrate as long as the heated vapour continues to fill it. And if it be wished to collect these volatile products of the wood, this can easily be done by introducing the neck of the retort into the water-bath apparatus, with which you are acquainted. But the preparation of common charcoal, such as is used in kitchens and manufactures, is performed on a much larger scale, and by an easier and less expensive process.

EMILY.

I have seen the process of making common charcoal. The wood is ranged on the ground in a pile of a pyramidical form, with a fire underneath; the whole is then covered with clay, a few holes only being left for the circulation of air.

MRS. B.

These holes are closed as soon as the wood is fairly lighted, so that the combustion is checked, or at least continues but in a very imperfect manner; but the heat produced by it is sufficient to force out and volatilize, through the earthy cover, most part of the oily and watery principles of the wood, although it cannot reduce it to ashes.

EMILY.

Is pure carbon as black as charcoal?

MRS. B.

The purest charcoal we can prepare is so; but chemists have never yet been able to separate it entirely from hydrogen. Sir H. Davy says, that the most perfect carbon that is prepared by art contains about five per cent. of hydrogen; he is of opinion, that if we could obtain it quite free from foreign ingredients, it would be metallic, in common with other simple substances.

But there is a form in which charcoal appears, that I dare say will surprise you.—This ring, which I wear on my finger, owes its brilliancy to a small piece of carbon.

CAROLINE.

Surely, you are jesting, Mrs. B.?

EMILY.

I thought your ring was diamond?

MRS. B.

It is so. But diamond is nothing more than carbon in a crystallized state.

EMILY.

That is astonishing! Is it possible to see two things apparently more different than diamond and charcoal?

CAROLINE.

It is, indeed, curious to think that we adorn ourselves with jewels of charcoal!

MRS. B.

There are many other substances, consisting chiefly of carbon, that are remarkably white. Cotton, for instance, is almost wholly carbon.

CAROLINE.

That, I own, I could never have imagined! — But pray, Mrs. B., since it is known of what substance diamond and cotton are composed, why should they not be manufactured, or imitated, by some chemical process, which would render them much cheaper, and more plentiful than the present mode of obtaining them?

MRS. B.

You might as well, my dear, propose that we should make flowers and fruit, nay, perhaps even animals, by a chemical process; for it is known of what these bodies consist, since every thing which we are acquainted with in nature is formed from the various simple substances that we have

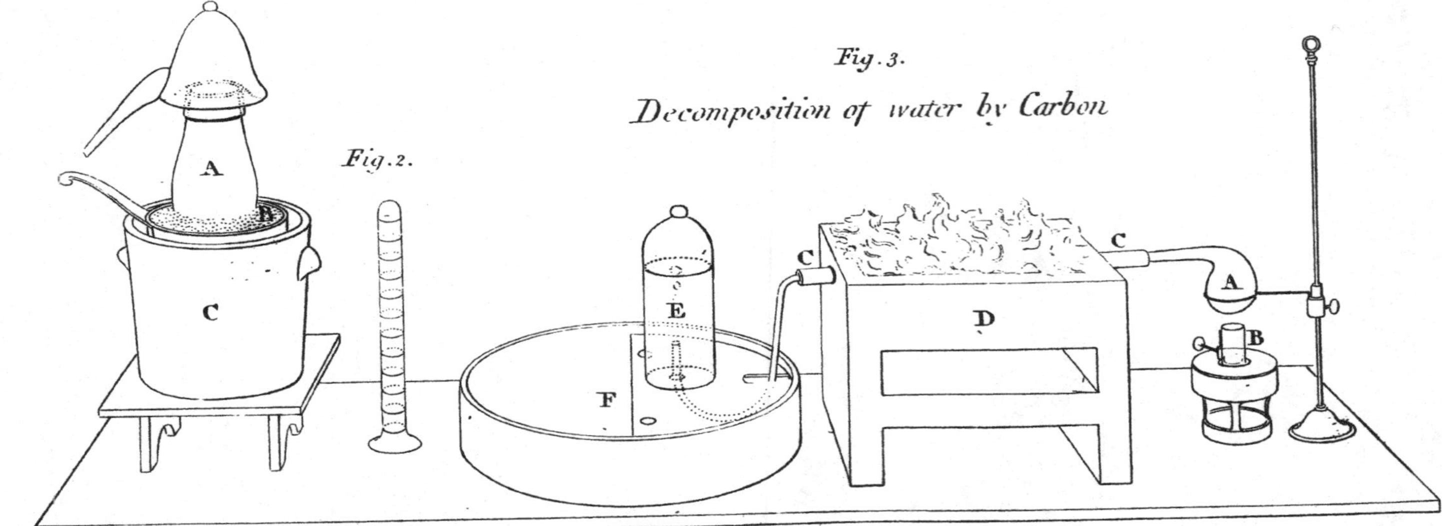

Fig. 1. A Alembic. _ B Sand-bath. _ C Furnace. _ Fig. 2. Eudiometer. _ Fig. 3. A Retort containing water. _ B Lamp to heat the water. _ C.C Porcelain tube containing Carbone. _ D Furnace through which the tube passes. _ E Receiver for the gas produced. _ F Water bath.

Drawn by the Author.

Published by Longman & Co. Oct.r 2nd 1809.

Engraved by Lowry.

enumerated. But you must not suppose that a knowledge of the component parts of a body will in every case enable us to imitate it. It is much less difficult to decompose bodies, and discover of what materials they are made, than it is to recompose them. The first of these processes is called *analysis*, the last *synthesis*. When we are able to ascertain the nature of a substance by both these methods, so that the result of one confirms that of the other, we obtain the most complete knowledge of it that we are capable of acquiring. This is the case with water, with the atmosphere, with most of the oxyds, acids, and neutral salts, and with many other compounds. But the more complicated combinations of nature, even in the mineral kingdom, are in general beyond our reach, and any attempt to imitate organised bodies must ever prove fruitless; their formation is a secret that rests in the bosom of the Creator. You see, therefore, how vain it would be to attempt to make cotton by chemical means. But, surely, we have no reason to regret our inability in this instance, when nature has so clearly pointed out a method of obtaining it in perfection and abundance.

CAROLINE.

I did not imagine that the principle of life could be imitated by the aid of chemistry; but it did not appear to me ridiculous to suppose that chemists might attain a perfect imitation of inanimate nature.

MRS. B.

They have succeeded in this point in a variety of instances; but, as you justly observe, the principle of life, or even the minute and intimate organisation of the vegetable kingdom, are secrets that have almost entirely eluded the researches of philosophers; nor do I imagine that human art will ever be capable of investigating them with complete success.

EMILY.

But diamond, since it consists of one simple unorganised substance, might be, one would think, perfectly imitable by art?

MRS. B.

It is sometimes as much beyond our power to obtain a simple body in a state of perfect purity, as it is to imitate a complicated combination; for the operations by which nature separates bodies are frequently as inimitable as those which she uses for their combination. This is the case with carbon; all the efforts of chemists to separate it entirely from other substances have been fruitless, and in the purest state in which it can be obtained by art, it still retains a portion of hydrogen, and probably of some other foreign ingredients. We are ignorant of the means which nature employs to crystallize it. It may probably be the work of ages, to purify, rrange, and unite the particles of carbon in the form of diamond. Here is some charcoal in the purest state we can procure it: you see that it is a very black, brittle, light, porous substance, entirely destitute of either taste or smell. Heat, without air, produces no alteration in it, as it is not volatile; but, on the contrary, it invariably remains at the bottom of the vessel after all the other parts of the vegetable are evaporated.

EMILY.

Yet carbon is, no doubt, combustible, since you say that charcoal would absorb oxygen if air were admitted during its preparation?

CAROLINE.

Unquestionably. Besides, you know, Emily, how much it is used in cooking. But pray what is the reason that charcoal burns without smoke, whilst a wood fire smokes so much?

MRS. B.

Because, in the conversion of wood into charcoal, the volatile particles of the former have been evaporated.

CAROLINE.

Yet I have frequently seen charcoal burn with

flame; therefore it must, in that case, contain some hydrogen.

MRS. B.

Very true; but you must recollect that charcoal, especially that which is used for common purposes, is not perfectly pure. It generally retains some remains of the various other component parts of vegetables, and hydrogen particularly, which accounts for the flame in question.

CAROLINE.

But what becomes of the carbon itself during its combustion?

MRS. B.

It gradually combines with the oxygen of the atmosphere, in the same way as sulphur and phosphorus, and, like those substances, it is converted into a peculiar acid, which flies off in a gaseous form. There is this difference, however, that the acid is not, in this instance, as in the two cases just mentioned, a mere condensable vapour, but a permanent elastic fluid, which always remains in the state of gas, under any pressure and at any temperature. The nature of this acid was first ascertained by Dr. Black, of Edinburgh; and, before the introduction of the new nomenclature, it was called *fixed air*. It is now distinguished by the more appropriate name of *carbonic acid gas*.

EMILY.

Carbon, then, can be volatilized by burning, though, by heat alone, no such effect is produced?

MRS. B.

Yes; but then it is no longer simple carbon, but an acid of which carbon forms the basis. In this state, carbon retains no more appearance of solidity or corporeal form, than the basis of any other gas. And you may, I think, from this instance, derive a more clear idea of the basis of the oxygen, hydrogen, and nitrogen gases, the existence of which, as real bodies, you seemed to doubt, because they were not to be obtained simply in a solid form.

EMILY.

That is true; we may conceive the basis of the oxygen, and of the other gases, to be solid, heavy substances, like carbon; but so much expanded by caloric as to become invisible.

CAROLINE.

But does not the carbonic acid gas partake of the blackness of charcoal?

MRS. B.

Not in the least. Blackness, you know, does not appear to be essential to carbon, and it is pure carbon, and not charcoal, that we must consider as the basis of carbonic acid. We shall make some carbonic acid, and, in order to hasten the process, we shall burn the carbon in oxygen gas.

EMILY.

But do you mean then to burn diamond?

MRS. B.

Charcoal will answer the purpose still better, being softer and more easy to inflame; besides the experiments on diamond are rather expensive.

CAROLINE.

But is it possible to burn diamond?

MRS. B.

Yes, it is; and in order to effect this combustion, nothing more is required than to apply a sufficient degree of heat by means of the blow-pipe, and of a stream of oxygen gas. Indeed it is by burning diamond that its chemical nature has been ascertained. It has long been known as a combustible substance, but it is within these few years only that the product of its combustion has been proved to be pure carbonic acid. This remarkable discovery is due to Mr. Tennant.

Now let us try to make some carbonic acid.— Will you, Emily, decant some oxygen gas from this large jar into the receiver in which we are to burn the carbon; and I shall introduce this small piece of charcoal, with a little lighted tinder, which will be necessary to give the first impulse to the combustion.

EMILY.

I cannot conceive how so small a piece of tinder, and that but just lighted, can raise the temperature of the carbon sufficiently to set fire to it; for it can produce scarcely any sensible heat, and it hardly touches the carbon.

MRS. B.

The tinder thus kindled has only heat enough to begin its own combustion, which, however, soon becomes so rapid in the oxygen gas, as to raise the temperature of the charcoal sufficiently for this to burn likewise, as you see is now the case.

EMILY.

I am surprised that the combustion of carbon is not more brilliant; it does not give out near so much light or caloric as phosphorus, or sulphur. Yet since it combines with so much oxygen, why is not a proportional quantity of light and heat disengaged from the decomposition of the oxygen gas, and the union of its electricity with that of the charcoal?

MRS. B.

It is not surprising that less light and heat should be liberated in this than in almost any other combustion, since the oxygen, instead of entering into a solid or liquid combination, as it does in the phosphoric and sulphuric acids, is employed in forming another elastic fluid; it therefore parts with less of its caloric.

EMILY.

True; and, on second consideration, it appears, on the contrary, surprising that the oxygen should, in its combination with carbon, retain a sufficient portion of caloric to maintain both substances in a gaseous state.

CAROLINE.

We may then judge of the degree of solidity in which oxygen is combined in a burnt body, by the quantity of caloric liberated during its combustion?

MRS. B.

Yes; provided that you take into the account the quantity of oxygen absorbed by the combustible body, and observe the proportion which the caloric bears to it.

CAROLINE.

But why should the water, after the combustion of carbon, rise in the receiver, since the gas within it retains an aëriform state?

MRS. B.

Because the carbonic acid gas is gradually absorbed by the water; and this effect would be promoted by shaking the receiver.

EMILY.

The charcoal is now extinguished, though it is not nearly consumed; it has such an extraordinary avidity for oxygen, I suppose, that the receiver did not contain enough to satisfy the whole.

MRS. B.

That is certainly the case; for if the combustion were performed in the exact proportions of 28 parts of carbon to 72 of oxygen, both these ingredients would disappear, and 100 parts of carbonic acid would be produced.

CAROLINE.

Carbonic acid must be a very strong acid, since it contains so great a proportion of oxygen?

MRS. B.

That is a very natural inference; yet it is erroneous. For the carbonic is the weakest of all the acids. The strength of an acid seems to depend upon the nature of its basis, and its mode of combination, as well as upon the proportion of the acidifying principle. The same quantity of oxygen that will convert some bodies into strong acids, will only be sufficient simply to oxydate others.

CAROLINE.

Since this acid is so weak, I think chemists should have called it the *carbonous*, instead of the *carbonic* acid.

EMILY.

But, I suppose, the carbonous acid is still weaker, and is formed by burning carbon in atmospherical air.

MRS. B.

It has been lately discovered, that carbon may be converted into a gas, by uniting with a smaller proportion of oxygen; but as this gas does not possess any acid properties, it is no more than an oxyd; it is called *gaseous oxyd of carbon.*

CAROLINE.

Pray is not carbonic acid a very wholesome gas to breathe, as it contains so much oxygen?

MRS. B.

On the contrary, it is extremely pernicious. Oxygen, when in a state of combination with other substances, loses, in almost every instance, its respirable properties, and the salubrious effects which it has on the animal economy when in its unconfined state. Carbonic acid is not only unfit for respiration, but extremely deleterious if taken into the lungs.

EMILY.

You know, Caroline, how very unwholesome the fumes of burning charcoal are reckoned.

CAROLINE.

Yes; but, to confess the truth, I did not consider that a charcoal fire produced carbonic acid gas. — Can this gas be condensed into a liquid?

MRS. B.

No: for, as I told you before, it is a permanent elastic fluid. But water can absorb a certain quantity of this gas, and can even be impregnated with it, in a very strong degree, by the assistance of agitation and pressure, as I am going to show you. I shall decant some carbonic acid gas into this bottle, which I fill first with water, in order to exclude the atmospherical air; the gas is then introduced through the water, which you see it displaces, for it will not mix with it in any quantity, unless strongly agitated, or allowed to stand over it for some time. The bottle is now about half full of carbonic acid gas, and the other half is still occupied by the water. By corking the bottle, and then violently shaking it, in this way, I can mix the gas and water together. — Now will you taste it?

EMILY.

It has a distinct acid taste.

CAROLINE.

Yes, it is sensibly sour, and appears full of little bubbles.

MRS. B.

It possesses likewise all the other properties of acids, but, of course, in a less degree than the pure carbonic acid gas, as it is so much diluted by water. This is a kind of artificial Seltzer water. By analysing that which is produced by nature, it was found to contain scarcely any thing more than common water impregnated with a certain proportion of carbonic acid gas. We are, therefore, able to imitate it, by mixing those proportions of water and carbonic acid. Here, my dear, is an instance, in which, by a chemical process, we can exactly copy the operations of nature; for the artificial Seltzer waters can be made in every respect similar to those of nature; in one point, indeed, the former have an advantage, since thay may be prepared stronger, or weaker, as occasion requires.

CAROLINE.

I thought I had tasted such water before. But what renders it so brisk and sparkling?

MRS. B.

This sparkling, or effervescence, as it is called, is always occasioned by the action of an elastic fluid escaping from a liquid; in the artifical Seltzer water, it is produced by the carbonic acid, which being lighter than the water in which it was strongly condensed, flies off with great rapidity the instant the bottle is uncorked; this makes it necessary to drink it immediately. The bubbling that took place in this bottle was but trifling, as the water was but very slightly impregnated with carbonic acid. It requires a particular apparatus to prepare the gaseous artificial mineral waters.

EMILY.

If, then, a bottle of Seltzer water remains for any length of time uncorked, I suppose it returns to the state of common water?

MRS. B.

The whole of the carbonic acid gas, or very nearly so, will soon disappear; but there is likewise in Seltzer water a very small quantity of soda, and of a few other saline or earthy ingredients, which will remain in the water, though it should be kept uncorked for any length of time.

CAROLINE.

I have often heard of people drinking soda-water. Pray what sort of water is that?

MRS. B.

It is a kind of artificial Seltzer water, holding

in solution, besides the gaseous acid, a particular saline substance, called soda, which imparts to the water certain medicinal qualities.

CAROLINE.

But how can these waters be so wholesome, since carbonic acid is so pernicious?

MRS. B.

A gas, we may conceive, though very prejudicial to breathe, may be beneficial to the stomach. — But it would be of no use to attempt explaining this more fully at present.

CAROLINE.

Are waters never impregnated with other gases?

MRS. B.

Yes; there are several kinds of gaseous waters. I forgot to tell you that waters have, for some years past, been prepared, impregnated both with oxygen and hydrogen gases. These are not an imitation of nature, but are altogether obtained by artificial means. They have been lately used medicinally, particularly on the continent, where, I understand, they have acquired some reputation.

EMILY.

If I recollect right, Mrs. B., you told us that

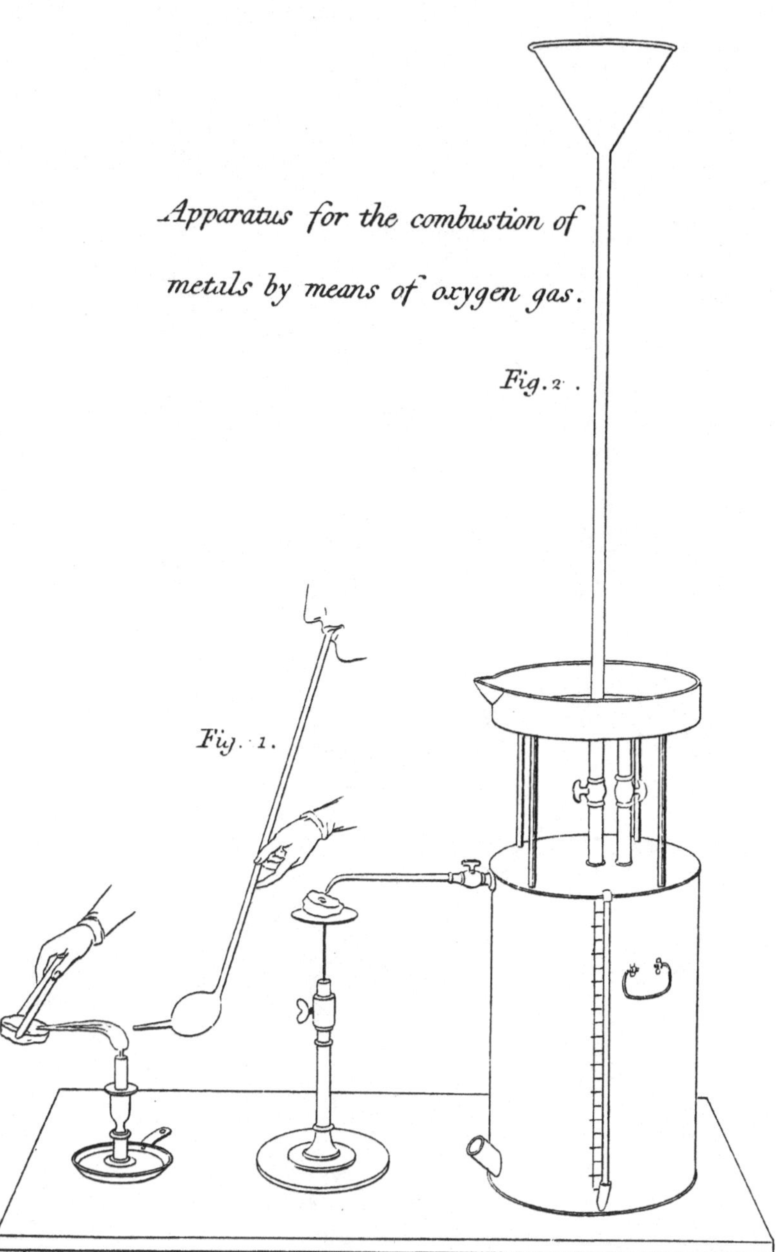

Apparatus for the combustion of metals by means of oxygen gas.
Fig. 2.
Fig. 1.

Fig. 1, Igniting charcoal with a taper & blow-pipe.—Fig. 2, Combustion of metals by means of a blow-pipe conveying a stream of oxygen gas from a gas holder.

Drawn by the Author. *Engraved by Lowry.*

Published by Longman & C°. Oct.r 2nd 1809.

carbon was capable of decomposing water; the affinity between oxygen and carbon must, therefore, be greater than between oxygen and hydrogen?

MRS. B.

Yes; but this is not the case unless their temperature be raised to a certain degree. It is only when carbon is red-hot, that it is capable of separating the oxygen from the hydrogen. Thus, if a small quantity of water be thrown on a red-hot fire, it will increase rather than extinguish the combustion; for the coals or wood (both of which contain a quantity of carbon) decompose the water, and thus supply the fire both with oxygen and hydrogen gases. If, on the contrary, a large mass of water be thrown over the fire, the diminution of heat thus produced is such, that the combustible matter loses the power of decomposing the water, and the fire is extinguished.

EMILY.

I have heard that fire-engines sometimes do more harm than good, and that they actually increase the fire when they cannot throw water enough to extinguish it. It must be owing, no doubt, to the decomposition of the water by the carbon during the conflagration.

MRS. B.

Certainly. — The apparatus which you see here (Plate XI fig. 3.), may be used to exemplify what we have just said. It consists in a kind of open furnace, through which a porcelain tube, containing charcoal, passes. To one end of the tube is adapted a glass retort with water in it; and the other end communicates with a receiver placed on the water-bath. A lamp being applied to the retort, and the water made to boil, the vapour is gradually conveyed through the red-hot charcoal, by which it is decomposed; and the hydrogen gas which results from this decomposition is collected in the receiver. But the hydrogen thus obtained is far from being pure; it retains in solution a minute portion of carbon, and contains also a quantity of carbonic acid. This renders it heavier than pure hydrogen gas, and gives it some peculiar properties; it is distinguished by the name of *carbonated hydrogen gas.*

CAROLINE.

And whence does it obtain the carbonic acid that is mixed with it?

EMILY.

I believe I can answer that question, Caroline. — From the union of the oxygen (proceeding from the decomposed water) with the carbon, which, you know, makes carbonic acid.

CAROLINE.

True; I should have recollected that. — The product of the decomposition of water by red-hot charcoal, therefore, is carbonated hydrogen gas, and carbonic acid gas.

MRS. B.

You are perfectly right now.

Carbon is frequently found combined with hydrogen in a state of solidity, especially in coals, which owe their combustible nature to these two principles.

EMILY.

Is it the hydrogen, then, that produces the flame of coals?

MRS. B.

It is so; and when all the hydrogen is consumed, the carbon continues to burn without flame. But again, as I mentioned when speaking of the gas-lights, the hydrogen gas produced by the burning of coals is not pure; for, during the combustion, particles of carbon are successively volatilized with the hydrogen, with which they form what is called a *hydro-carbonat*, which is the principal product of this combustion.

Carbon is a very bad conductor of heat; for

this reason, it is employed (in conjunction with other ingredients) for coating furnaces and other chemical apparatus.

EMILY.

Pray what is the use of coating furnaces?

MRS. B.

In most cases, in which a furnace is used, it is necessary to produce and preserve a great degree of heat, for which purpose every possible means are used to prevent the heat from escaping by communicating with other bodies, and this object is attained by coating over the inside of the furnace with a kind of plaster, composed of materials that are bad conductors of heat.

Carbon, combined with a small quantity of iron, forms a compound called plumbago, or black-lead, of which pencils are made. This substance, agreeably to the nomenclature, is *a carburet of iron.*

EMILY.

Why, then, is it called black-lead?

MRS. B.

It is an ancient name given to it by ignorant people, from its shining metallic appearance; but it is certainly a most improper name for it, as there is not a particle of lead in the composition.

There is only one mine of this mineral, which is in Cumberland. It is supposed to approach as nearly to pure carbon as the best prepared charcoal does, as it contains only five parts of iron, unadulterated by any other foreign ingredients. There is another carburet of iron, in which the iron, though united only to an extremely small proportion of carbon, acquires very remarkable properties; this is steel.

CAROLINE.

Really; and yet steel is much harder than iron?

MRS. B.

But carbon is not ductile like iron, and therefore may render the steel more brittle, and prevent its bending so easily. Whether it is that the carbon, by introducing itself into the pores of the iron, and, by filling them, makes the metal both harder and heavier; or whether this change depends upon some chemical cause, I cannot pretend to decide. But there is a subsequent operation, by which the hardness of steel is very much increased, which simply consists in heating the steel till it is red-hot, and then plunging it into cold water.

Carbon, besides the combination just mentioned, enters into the composition of a vast number of natural productions, such, for instance, as all the various kinds of oils, which result from the combination of carbon, hydrogen, and caloric, in various proportions.

EMILY.

I thought that carbon, hydrogen, and caloric, formed carbonated hydrogen gas?

MRS. B.

That is the case when a small portion of carbonic acid gas is held in solution by hydrogen gas. Different proportions of the same principles, together with the circumstances of their union, produce very different combinations; of this you will see innumerable examples. Besides, we are not now talking of gases, but of carbon and hydrogen, combined only with a quantity of caloric sufficient to bring them to the consistency of oil or fat.

CAROLINE.

But oil and fat are not of the same consistence?

MRS. B.

Fat is only congealed oil; or oil, melted fat. The one requires a little more heat to maintain it in a fluid state than the other. Have you never observed the fat of meat turned to oil by the caloric it has imbibed from the fire?

EMILY.

Yet oils in general, as salad-oil, and lamp-oil, do not turn to fat when cold?

MRS. B.

Not at the common temperature of the atmosphere, because they retain too much caloric to congeal at that temperature; but if exposed to a sufficient degree of cold, their latent heat is extricated, and they become solid fat substances. Have you never seen salad oil frozen in winter?

EMILY.

Yes; but it appears to me in that state very different from animal fat.

MRS. B.

The essential constituent parts of either vegetable or animal oils are the same, carbon and hydrogen; their variety arises from the different proportions of these substances, and from other accessory ingredients that may be mixed with them. The oil of a whale, and the oil of roses, are, in their essential constituent parts, the same; but the one is impregnated with the offensive particles of animal matter, the other with the delicate perfume of a flower.

The difference of *fixed oils*, and *volatile* or *essential oils*, consists also in the various proportions of carbon and hydrogen. Fixed oils are those which will not evaporate without being decomposed; this is the case with all common oils, which contain a greater proportion of carbon than the essential oils. The essential oils (which comprehend the whole class of essences and perfumes) are lighter; they contain more equal proportions of carbon and hydrogen, and are volatilized or evaporated without being decomposed.

EMILY.

When you say that one kind of oil will evaporate, and the other be decomposed, you mean, I suppose, by the application of heat?

MRS. B.

Not necessarily; for there are oils that will evaporate slowly at the common temperature of the atmosphere; but for a more rapid volatilization, or for their decomposition, the assistance of heat is required.

CAROLINE.

I shall now remember, I think, that fat and oil are really the same substances, both consisting of carbon and hydrogen; that in fixed oils the carbon preponderates, and heat produces a decomposition; while, in essential oils, the proportion of hydrogen is greater, and heat produces a volatilization only.

EMILY.

I suppose the reason why oil burns so well in

lamps is because its two constituents are so combustible?

MRS. B.

Certainly; the combustion of oil is just the same as that of a candle; if tallow, it is only oil in a concrete state; if wax, or spermaceti, its chief chemical ingredients are still hydrogen and carbon.

EMILY.

I wonder, then, there should be so great a difference between tallow and wax?

MRS. B.

I must again repeat, that the same substances, in different proportions, produce results that have sometimes scarcely any resemblance to each other. But this is rather a general remark that I wish to impress upon your minds, than one which is applicable to the present case; for tallow and wax are far from being very dissimilar; the chief difference consists in the wax being a purer compound of carbon and hydrogen than the tallow, which retains more of the gross particles of animal matter. The combustion of a candle, and that of a lamp, both produce water and carbonic acid gas. Can you tell me how these are formed?

EMILY.

Let me reflect . . . . Both the candle and lamp

burn by means of fixed oil — this is decomposed as the combustion goes on; and the constituent parts of the oil being thus separated, the carbon unites to a portion of oxygen from the atmosphere to form carbonic acid gas, whilst the hydrogen combines with another portion of oxygen, and forms with it water. — The products, therefore, of the combustion of oils are water and carbonic acid gas.

CAROLINE.

But we see neither water nor carbonic acid produced by the combustion of a candle.

MRS. B.

The carbonic acid gas, you know, is invisible, and the water being in a state of vapour, is so likewise. Emily is perfectly correct in her explanation, and I am very much pleased with it.

All the vegetable acids consist of various proportions of carbon and hydrogen, acidified by oxygen. Gums, sugar, and starch, are likewise composed of these ingredients; but, as the oxygen which they contain is not sufficient to convert them into acids, they are classed with the oxyds, and called vegetable oxyds.

CAROLINE.

I am very much delighted with all these new

ideas; but, at the same time, I cannot help being apprehensive that I may forget many of them.

MRS. B.

I would advise you to take notes, or, what would answer better still, to write down, after every lesson, as much of it as you can recollect. And, in order to give you a little assistance, I shall lend you the heads or index, which I occasionally consult for the sake of preserving some method and arrangement in these conversations. Unless you follow some such plan, you cannot expect to retain nearly all that you learn, how great soever be the impression it may make on you at first.

EMILY.

I will certainly follow your advice. — Hitherto I have found that I recollected pretty well what you have taught us; but the history of carbon is a more extensive subject than any of the simple bodies we have yet examined.

MRS. B.

I have little more to say on carbon at present; but hereafter you will see that it performs a considerable part in most chemical operations.

CAROLINE.

That is, I suppose, owing to its entering into

the composition of so great a variety of substances?

MRS. B.

Certainly; it is the basis, you have seen, of all vegetable matter; and you will find that it is very essential to the process of animalization. But in the mineral kingdom also, particularly in its form of carbonic acid, we shall often discover it combined with a great variety of substances.

In chemical operations, carbon is particularly useful, from its very great attraction for oxygen, as it will absorb this substance from many oxygenated or burnt bodies, and thus deoxygenate, or *unburn* them, and restore them to their original combustible state.

CAROLINE.

I do not understand how a body can be *unburnt*, and restored to its original state. This piece of tinder, for instance, that has been burnt, if by any means the oxygen were extracted from it, would not be restored to its former state of linen; for its texture is destroyed by burning, and that must be the case with all organized or manufactured substances, as you observed in a former conversation.

MRS. B.

A compound body is decomposed by combustion in a way which generally precludes the possibility of restoring it to its former state; the oxygen, for instance, does not become fixed in the tinder, but it combines with its volatile parts, and flies off in the shape of gas, or watery vapour. You see, therefore, how vain it would be to attempt the recomposition of such bodies. But, with regard to simple bodies, or at least bodies whose component parts are not disturbed by the process of oxygenation or deoxygenation, it is often possible to restore them, after combustion, to their original state. — The metals, for instance, undergo no other alteration by combustion than a combination with oxygen; therefore, when the oxygen is taken from them, they return to their pure metallic state. But I shall say nothing further of this at present, as the metals will furnish ample subject for another morning; and they are the class of simple bodies that come next under consideration.

CONVERSATION X.

ON METALS.

MRS. B.

THE METALS, which we are now to examine, are bodies of a very different nature from those which we have hitherto considered. They do not, like the bases of gases, elude the immediate observation of our senses; for they are the most brilliant, the most ponderous, and the most palpable substances in nature.

CAROLINE.

I doubt, however, whether the metals will appear to us so interesting, and give us so much entertainment as those mysterious elements which conceal themselves from our view. Besides, they cannot afford so much novelty; they are bodies with which we are already so well acquainted.

MRS. B.

You are not aware, my dear, of the interesting discoveries which were a few years ago made by Sir H. Davy respecting this class of bodies. By the aid of the Voltaic battery, he has obtained from a variety of substances, metals before unknown, the properties of which are equally new and curious. We shall begin, however, by noticing those metals with which you profess to be so well acquainted. But the acquaintance, you will soon perceive, is but very superficial; and I trust that you will find both novelty and entertainment in considering the metals in a chemical point of view. To treat of this subject fully, would require a whole course of lectures; for metals form of themselves a most important branch of practical chemistry. We must, therefore, confine ourselves to a general view of them. These bodies are seldom found naturally in their metallic form: they are generally more or less oxygenated or combined with sulphur, earths, or acids, and are often blended with each other. They are found buried in the bowels of the earth in most parts of the world, but chiefly in mountainous districts, where the surface of the globe has suffered from the earthquakes, volcanos, and other convulsions of nature. They are spread in strata or beds, called veins, and these veins are composed of a certain quantity of metal, combined with various earthy substances, with which they form minerals of different nature and appearance, which are called *ores*.

CAROLINE.

I now feel quite at home, for my father has

a lead-mine in Yorkshire, and I have heard a great deal about veins of ore, and of the *roasting* and *smelting* of the lead; but, I confess, that I do not understand in what these operations consist.

MRS. B.

Roasting is the process by which the volatile parts of the ore are evaporated; smelting, that by which the pure metal is afterwards separated from the earthy remains of the ore. This is done by throwing the whole into a furnace, and mixing with it certain substances that will combine with the earthy parts and other foreign ingredients of the ore; the metal being the heaviest, falls to the bottom, and runs out by proper openings in its pure metallic state.

EMILY.

You told us in a preceding lesson that metals had a great affinity for oxygen. Do they not, therefore, combine with oxygen, when strongly heated in the furnace, and run out in the state of oxyds?

MRS. B.

No; for the scoriæ, or oxyd, which soon forms on the surface of the fused metal, when it is oxydable, prevents the air from having any further influence on the mass; so that neither combustion nor oxygenation can take place.

CAROLINE.

Are all the metals equally combustible?

MRS. B.

No; their attraction for oxygen varies extremely. There are some that will combine with it only at a very high temperature, or by the assistance of acids; whilst there are others that oxydate spontaneously and with great rapidity, even at the lowest temperature; such is in particular manganese, which scarcely ever exists in the metallic state, as it immediately absorbs oxygen on being exposed to the air, and crumbles to an oxyd in the course of a few hours.

EMILY.

Is not that the oxyd from which you extracted the oxygen gas?

MRS. B.

It is: so that, you see, this metal attracts oxygen at a low temperature, and parts with it when strongly heated.

EMILY.

Is there any other metal that oxydates at the temperature of the atmosphere?

MRS. B.

They all do, more or less, excepting gold, silver, and platina.

Copper, lead, and iron, oxydate slowly in the air, and cover themselves with a sort of rust, a process which depends on the gradual conversion of the surface into an oxyd. This rusty surface preserves the interior metal from oxydation, as it prevents the air from coming in contact with it. Strictly speaking, however, the word rust applies only to the oxyd, which forms on the surface of iron, when exposed to air and moisture, which oxyd appears to be united with a small portion of carbonic acid.

EMILY.

When metals oxydate from the atmosphere without an elevation of temperature, some light and heat, I suppose, must be disengaged, though not in sufficient quantities to be sensible.

MRS. B.

Undoubtedly; and, indeed, it is not surprising that in this case the light and heat should not be sensible, when you consider how extremely slow, and, indeed, how imperfectly, most metals oxydate by mere exposure to the atmosphere. For the quantity of oxygen with which metals are capable of combining, generally depends upon their temperature; and the absorption stops at various points of oxydation, according to the degree to which their temperature is raised.

EMILY.

That seems very natural; for the greater the quantity of caloric introduced into a metal, the more will its positive electricity be exalted, and consequently the stronger will be its affinity for oxygen.

MRS. B.

Certainly. When the metal oxygenates with sufficient rapidity for light and heat to become sensible, combustion actually takes place. But this happens only at very high temperatures, and the product is nevertheless an oxyd; for though, as I have just said, metals will combine with different proportions of oxygen, yet with the exception of only five of them, they are not susceptible of acidification.

Metals change colour during the different degrees of oxydation which they undergo. Lead, when heated in contact with the atmosphere, first becomes grey; if its temperature be then raised, it turns yellow, and a still stronger heat changes it to red. Iron becomes successively a green, brown, and white oxyd. Copper changes from brown to blue, and lastly green.

EMILY.

Pray, is the white lead with which houses are painted prepared by oxydating lead?

MRS. B.

Not merely by oxydating, but by being also united with carbonic acid. It is a carbonat of lead. The mere oxyd of lead is called red lead. Litharge is another oxyd of lead, containing less oxygen. Almost all the metallic oxyds are used as paints. The various sorts of ochres consist chiefly of iron more or less oxydated. And it is a remarkable circumstance, that if you burn metals rapidly, the light or flame they emit during combustion partakes of the colours which the oxyd successively assumes.

CAROLINE.

How is that accounted for, Mrs. B.? For light, you know, does not proceed from the burning body, but from the decomposition of the oxygen gas?

MRS. B.

The correspondence of the colour of the light with that of the oxyd which emits it, is, in all probability, owing to some particles of the metal which are volatilised and carried off by the caloric.

CAROLINE.

It is then a sort of metallic gas.

EMILY.

Why is it reckoned so unwholesome to breathe the air of a place in which metals are melting?

MRS. B.

Perhaps the notion is too generally entertained. But it is true with respect to lead, and some other noxious metals, because, unless care be taken, the particles of the oxyd which are volatilised by the heat are inhaled in with the breath, and may produce dangerous effects.

I must show you some instances of the combustion of metals; it would require the heat of a furnace to make them burn in the common air, but if we supply them with a stream of oxygen gas, we may easily accomplish it.

CAROLINE.

But it will still, I suppose, be necessary in some degree to raise their temperature?

MRS. B.

This, as you shall see, is very easily done, particularly if the experiment be tried upon a small scale. — I begin by lighting this piece of charcoal with the candle, and then increase the rapidity of of its combustion by blowing upon it with a blow-pipe. (Plate XII. fig. 1.)

EMILY.

That I do not understand; for it is not every kind of air, but merely oxygen gas, that produces combustion. Now you said that in breathing we inspired, but did not expire oxygen gas. Why, therefore, should the air which you breathe through the blow-pipe promote the combustion of the charcoal?

MRS. B.

Because the air, which has but once passed through the lungs, is yet but little altered, a small portion only of its oxygen being destroyed; so that a great deal more is gained by increasing the rapidity of the current, by means of the blow-pipe, than is lost in consequence of the air passing once through the lungs, as you shall see —

EMILY.

Yes, indeed, it makes the charcoal burn much brighter.

MRS. B.

Whilst it is red-hot, I shall drop some iron filings on it, and supply them with a current of oxygen gas, by means of this apparatus, (Plate XII. fig. 2.) which consists simply of a closed tin cylindrical vessel, full of oxygen gas, with two apertures and stop-cocks, by one of which a stream of water is thrown into the vessel through a long funnel, whilst by the other the gas is forced out through a blow-pipe adapted to it, as the water gains admittance. — Now that I pour water into the funnel, you may hear the gas issuing from the blow-pipe — I bring the charcoal close to the current, and drop the filings upon it —

CAROLINE.

They emit much the same vivid light as the combustion of the iron wire in oxygen gas.

MRS. B.

The process is, in fact, the same; there is only some difference in the mode of conducting it. Let us burn some tin in the same manner — you see that it is equally combustible. — Let us now try some copper —

CAROLINE.

This burns with a greenish flame; it is, I suppose, owing to the colour of the oxyd?

EMILY.

Pray, shall we not also burn some gold?

MRS. B.

That is not in our power, at least in this way. Gold, silver, and platina, are incapable of being oxydated by the greatest heat that we can produce by the common method. It is from this circumstance, that they have been called perfect metals. Even these, however, have an affinity for oxygen; but their oxydation or combustion can be performed only by means of acids or by electricity.

The spark given out by the Voltaic battery produces at the point of contact a greater degree of heat than any other process; and it is at this very high temperature only that the affinity of these metals for oxygen will enable them to act on each other.

I am sorry that I cannot show you the combustion of the perfect metals by this process, but it requires a considerable Voltaic battery. You will see these experiments performed in the most perfect manner, when you attend the chemical lectures of the Royal Institution. But in the mean time I can, without difficulty, show you an ingenious apparatus lately contrived for the purpose of producing intense heats, the power of which nearly equals that of the largest Voltaic batteries. It simply consists, you see, in a strong box, made of iron or copper, (PLATE X. fig. 2.) to which may be adapted this air-syringe or condensing-pump, and a stop-cock terminating in a small orifice similar to that of a blow-pipe. By working the condensing syringe, up and down in this manner, a quantity of air is accumulated in the vessel, which may be increased to almost any extent; so that if we now turn the stop-cock, the condensed air will rush out, forming a jet of considerable force; and if we place the flame of a lamp in the current, you will see how violently the flame is driven in that direction.

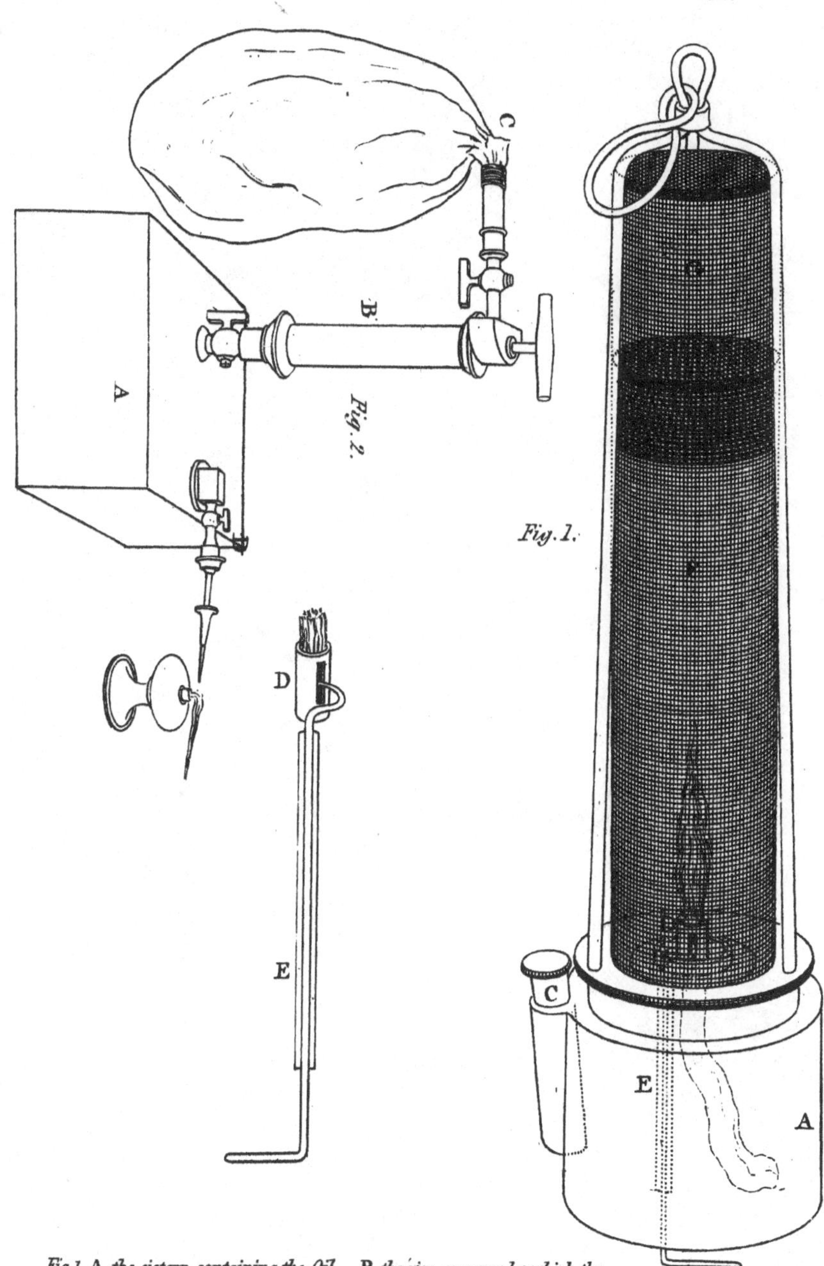

Fig 1. A. *the cistern containing the Oil.* — B. *the rim or screw by which the gauze cage is fixed to the cistern.* — C, *apperture for supplying Oil.* E. *a wire for trimming the wick.* D. — F. *the wire gauze cylinder.* — G. *a double top.*

Lowry

Fig. 2. A. *the reservoir of condensed air.* — B. *the condensing Syringe.* C. *the bladder for Oxygen.* — D. *the moveable jet.*

Published by Longman & C° May 1st 1817.

CAROLINE.

It seems to be exactly the same effect as that of a blow-pipe worked by the mouth, only much stronger.

EMILY.

Yes; and this new instrument has this additional advantage, that it does not fatigue the mouth and lungs like the common blow-pipe, and requires no art in blowing.

MRS. B.

Unquestionably; but yet this blow-pipe would be of very limited utility, if its energy and power could not be greatly increased by some other contrivance. Can you imagine any mode of producing such an effect?

EMILY.

Could not the reservoir be charged with pure oxygen, instead of common air, as in the case of the gas-holder?

MRS. B.

Undoubtedly; and this is precisely the contrivance I allude to. The vessel need only be supplied with air from a bladder full of oxygen, instead of the air of the room, and this, you see, may be easily done by screwing the bladder on the upper part of the syringe, so that in working the syringe the oxygen gas is forced from the bladder into the condensing vessel.

CAROLINE.

With the aid of this small apparatus, therefore, we could obtain the same effects as those we have just produced with the gas-holder, by means of a column of water forcing the gas out of it?

MRS. B.

Yes; and much more conveniently so. But there is a mode of using this apparatus by which more powerful effects still may be obtained. It consists in condensing in the reservoir, not oxygen alone, but a mixture of oxygen and hydrogen in the exact proportion in which they unite to produce water; and then kindling the jet formed by the mixed gases. The heat disengaged by this combustion, without the help of any lamp, is probably the most intense known; and various effects are said to have been obtained from it which exceed all expectation.

CAROLINE.

But why should we not try this experiment?

MRS. B.

Because it is not exempt from danger; the combustion (notwithstanding various contrivances which have been resorted to with a view to prevent accident) being apt to penetrate into the inside of the vessel, and to produce a dangerous and violent explosion.—We shall, therefore, now proceed in our subject.

CAROLINE.

I think you said the oxyds of metals could be restored to their metallic state?

MRS. B.

Yes; this is called *reviving* a metal. Metals are in general capable of being revived by charcoal, when heated red hot, charcoal having a greater attraction for oxygen than the metals. You need only, therefore, decompose, or unburn the oxyd, by depriving it of its oxygen, and the metal will be restored to its pure state.

EMILY.

But will the carbon, by this operation, be burnt, and be converted into carbonic acid?

MRS. B.

Certainly. There are other combustible substances to which metals at a high temperature will part with their oxygen. They will also yield it to each other, according to their several degrees of attraction for it; and if the oxygen goes into a more dense state in the metal which it enters, than it existed in that which it quits, a proportional disengagement of caloric will take place.

CAROLINE.

And cannot the oxyds of gold, silver, and platina, which are formed by means of acids or of the electric fluid, be restored to their metallic state?

MRS. B.

Yes, they may; and the intervention of a combustible body is not required; heat alone will take the oxygen from them, convert it into a gas, and revive the metal.

EMILY.

You said that rust was an oxyd of iron; how is it, then, that water, or merely dampness, produces it, which, you know, it very frequently does on steel grates, or any iron instruments?

MRS. B.

In that case the metal decomposes the water, or dampness (which is nothing but water in a state of vapour), and obtains the oxygen from it.

CAROLINE.

I thought that it was necessary to bring metals to a very high temperature to enable them to decompose water.

MRS. B.

It is so, if it is required that the process should be performed rapidly, and if any considerable quantity is to be decomposed. Rust, you know, is sometimes months in forming, and then it is only the surface of the metal that is oxydated.

EMILY.

Metals, then, that do not rust, are incapable of spontaneous oxydation, either by air or water?

MRS. B.

Yes; and this is the case with the perfect metals, which, on that account, preserve their metallic lustre so well.

EMILY.

Are all metals capable of decomposing water, provided their temperature be sufficiently raised?

MRS. B.

No; a certain degree of attraction is requisite, besides the assistance of heat. Water, you recollect, is composed of oxygen and hydrogen; and, unless the affinity of the metal for oxygen be stronger than that of hydrogen, it is in vain that we raise its temperature, for it cannot take the oxygen from the hydrogen. Iron, zinc, tin, and antimony, have a stronger affinity for oxygen than hydrogen has, therefore these four metals are capable of decomposing water. But hydrogen having an advantage over all the other metals with respect to its affinity for oxygen, it not only withholds its oxygen from them, but is even capable under certain circumstances, of taking the oxygen from the oxyds of these metals.

EMILY.

I confess that I do not quite understand why hydrogen can take oxygen from those metals that do not decompose water.

CAROLINE.

Now I think I do perfectly. Lead, for instance, will not decompose water, because it has not so strong an attraction for oxygen as hydrogen has. Well, then, suppose the lead to be in a state of oxyd; hydrogen will take the oxygen from the lead, and unite with it to form water, because hydrogen has a stronger attraction for oxygen, than oxygen has for lead; and it is the same with all the other metals which do not decompose water.

EMILY.

I understand your explanation, Caroline, very well; and I imagine that it is because lead cannot decompose water that it is so much employed for pipes for conveying that fluid.

MRS. B.

Certainly; lead is, on that account, particularly appropriate to such purposes; whilst, on the contrary, this metal, if it was oxydable by water, would impart to it very noxious qualities, as all oxyds of lead are more or less pernicious.

But, with regard to the oxydation of metals, the most powerful mode of effecting it is by means of acids. These, you know, contain a much greater proportion of oxygen than either air or water; and will, most of them, easily yield it to metals. Thus, you recollect, the zinc plates of the Voltaic battery are oxydated by the acid and water, much more effectually than by water alone.

CAROLINE.

And I have often observed that if I drop vinegar, lemon, or any acid on the blade of a knife, or on a pair of scissars, it will immediately produce a spot of rust.

EMILY.

Metals have, then, three ways of obtaining oxygen; from the atmosphere, from water, and from acids.

MRS. B.

The two first you have already witnessed, and I shall now show you how metals take the oxygen from an acid. This bottle contains nitric acid; I shall pour some of it over this piece of copper-leaf . . . .

CAROLINE.

Oh, what a disagreeable smell!

EMILY.

And what is it that produces the effervescence and that thick yellow vapour?

MRS. B.

It is the acid, which being abandoned by the greatest part of its oxygen, is converted into a weaker acid, which escapes in the form of gas.

CAROLINE.

And whence proceeds this heat?

MRS. B.

Indeed, Caroline, I think you might now be able to answer that question yourself.

CAROLINE.

Perhaps it is that the oxygen enters into the metal in a more solid state than it existed in the acid, in consequence of which caloric is disengaged.

MRS. B.

If the combination of the oxygen and the metal results from the union of their opposite electricities, of course caloric must be given out.

EMILY.

The effervescence is over; therefore I suppose that the metal is now oxydated.

MRS. B.

Yes. But there is another important connection between metals and acids, with which I must now make you acquainted. Metals, when in the state of oxyds, are capable of being dissolved by acids. In this operation they enter into a chemical combination with the acid, and form an entirely new compound.

CAROLINE.

But what difference is there between the *oxydation* and the *dissolution* of the metal by an acid?

MRS. B.

In the first case, the metal merely combines with a portion of oxygen taken from the acid, which is thus partly deoxygenated, as in the instance you have just seen; in the second case, the metal, after being previously oxydated, is actually dissolved in the acid, and enters into a chemical combination with it, without producing any further decomposition or effervescence. — This complete combination of an oxyd and an acid forms a peculiar and important class of compound salts.

EMILY.

The difference between an oxyd and a compound salt, therefore, is very obvious: the one consists of a metal and oxygen; the other of an oxyd and an acid.

MRS. B.

Very well: and you will be careful to remember that the metals are incapable of entering into this combination with acids, unless they are previously oxydated; therefore, whenever you bring a metal in contact with an acid, it will be first oxydated and afterwards dissolved, provided that there be a sufficient quantity of acid for both operations.

There are some metals, however, whose solution is more easily accomplished, by diluting the acid in water; and the metal will, in this case, be oxydated, not by the acid, but by the water, which it will decompose. But in proportion as the oxygen of the water oxydates the surface of the metal, the acid combines with it, washes it off, and leaves a fresh surface for the oxygen to act upon: then other coats of oxyd are successively formed, and rapidly dissolved by the acid, which continues combining with the new-formed surfaces of oxyd till the whole of the metal is dissolved. During this process the hydrogen gas of the water is disengaged, and flies off with effervescence.

EMILY.

Was not this the manner in which the sulphuric acid assisted the iron filings in decomposing water?

MRS. B.

Exactly; and it is thus that several metals, which are incapable alone of decomposing water, are enabled to do it by the assistance of an acid, which, by continually washing off the covering of oxyd, as it is formed, prepares a fresh surface of metal to act upon the water.

CAROLINE.

The acid here seems to act a part not very different from that of a scrubbing-brush. — But pray would not this be a good method of cleaning metallic utensils?

MRS. B.

Yes; on some occasions a weak acid, as vinegar, is used for cleaning copper. Iron plates, too, are freed from the rust on their surface by diluted muriatic acid, previous to their being covered with tin. You must remember, however, that in this mode of cleaning metals the acid should be quickly afterwards wiped off, otherwise it would produce fresh oxyd.

CAROLINE.

Let us watch the dissolution of the copper in the nitric acid; for I am very impatient to see the salt that is to result from it. The mixture is now of a beautiful blue colour; but there is no appearance of the formation of a salt; it seems to be a tedious operation.

MRS. B.

The crystallisation of the salt requires some length of time to be completed; if, however, you are so impatient, I can easily show you a metallic salt already formed.

CAROLINE.

But that would not satisfy my curiosity half so well as one of our own manufacturing.

MRS. B.

It is one of our own preparing that I mean to show you. When we decomposed water a few days since, by the oxydation of iron filings through the assistance of sulphuric acid, in what did the process consist?

CAROLINE.

In proportion as the water yielded its oxygen to the iron, the acid combined with the new-formed oxyd, and the hydrogen escaped alone.

MRS. B.

Very well; the result, therefore, was a compound salt, formed by the combination of sulphuric acid with oxyd of iron. It still remains in the vessel in which the experiment was performed. Fetch it, and we shall examine it.

EMILY.

What a variety of processes the decomposition of water, by a metal and an acid, implies; 1st, the decomposition of the water; 2dly, the oxydation of the metal; and 3dly, the formation of a compound salt.

CAROLINE.

Here it is, Mrs. B.—What beautiful green crystals! But we do not perceive any crystals in the solution of copper in nitrous acid?

MRS. B.

Because the salt is now suspended in the water which the nitrous acid contains, and will remain so till it is deposited in consequence of rest and cooling.

EMILY.

I am surprised that a body so opake as iron can be converted into such transparent crystals.

MRS. B.

It is the union with the acid that produces the transparency; for if the pure metal were melted, and afterwards permitted to cool and crystallise, it would be found just as opake as before.

EMILY.

I do not understand the exact meaning of *crystallisation?*

MRS. B.

You recollect that when a solid body is dissolved either by water or caloric it is not decomposed; but that its integrant parts are only suspended in the solvent. When the solution is made in water, the integrant particles of the body will, on the water being evaporated, again unite into a solid mass by the force of their mutual attraction. But when the body is dissolved by caloric alone, nothing more is necessary, in order to make its particles reunite, than to reduce its temperature. And, in general, if the solvent, whether water or caloric, be slowly separated by evaporation or by cooling, and care taken that the particles be not agitated during their reunion, they will arrange themselves in regular masses, each individual substance assuming a peculiar form or arrangement; and this is what is called crystallisation.

EMILY.

Crystallisation, therefore, is simply the reunion of the particles of a solid body that has been dissolved in a fluid.

MRS. B.

That is a very good definition of it. But I must not forget to observe, that *heat* and *water* may unite their solvent powers; and, in this case, crystallisation may be hastened by cooling, as well as by evaporating the liquid?

CAROLINE.

But if the body dissolved is of a volatile nature, will it not evaporate with the fluid?

MRS. B.

A crystallised body held in solution only by water is scarcely ever so volatile as the fluid itself, and care must be taken to manage the heat so that it may be sufficient to evaporate the water only.

I should not omit also to mention that bodies, in crystallising from their watery solution, always retain a small portion of water, which remains confined in the crystal in a solid form, and does not reappear unless the body loses its crystalline state. This is called the *water of crystallisation.* But you must observe, that whilst a body may be separated from its solution in water or caloric simply by cooling or by evaporation, an acid can be taken from a metal with which it is combined only by stronger affinities, which produce a decomposition.

EMILY.

Are the perfect metals susceptible of being dissolved and converted into compound salts by acids?

MRS. B.

Gold is acted upon by only one acid, the *oxygenated muriatic*, a very remarkable acid, which, when in its most concentrated state, dissolves gold or any other metal, by burning them rapidly.

Gold can, it is true, be dissolved likewise by a mixture of two acids, commonly called *aqua regia*; but this mixed solvent derives that property from containing the peculiar acid which I have just mentioned. Platina is also acted upon by this acid only; silver is dissolved by nitric acid.

CAROLINE.

I think you said that some of the metals might be so strongly oxydated as to become acid?

MRS. B.

There are five metals, arsenic, molybdena, chrome, tungsten, and columbium, which are susceptible of combining with a sufficient quantity of oxygen to be converted into acids.

CAROLINE.

Acids are connected with metals in such a variety of ways, that I am afraid of some confusion in remembering them. — In the first place, acids will yield their oxygen to metals. Secondly, they will combine with them in their state of oxyds, to form compound salts; and lastly, several of the metals are themselves susceptible of acidification.

MRS. B.

Very well; but though metals have so great an affinity for acids, it is not with that class of bodies alone that they will combine. They are most of them, in their simple state, capable of uniting with sulphur, with phosphorus, with carbon, and with each other; these combinations, according to the nomenclature which was explained to you on a former occasion, are called *sulphurets, phosphorets, carburets*, &c.

The metallic phosphorets offer nothing very remarkable. The sulphurets form the peculiar kind of mineral called *pyrites*, from which certain kinds of mineral waters, as those of Harrogate, derive their chief chemical properties. In this combination, the sulphur, together with the iron, have so strong an attraction for oxygen, that they obtain it both from the air and from water, and by condensing it in a solid form, produce the heat which raises the temperature of the water in such a remarkable degree.

EMILY.

But if pyrites obtain oxygen from water, that

water must suffer a decomposition, and hydrogen gas be evolved.

MRS. B.

That is actually the case in the hot springs alluded to, which give out an extremely fetid gas, composed of hydrogen impregnated with sulphur.

CAROLINE.

If I recollect right, steel and plumbago, which you mentioned in the last lesson, are both carburets of iron?

MRS. B.

Yes; and they are the only carburets of much consequence.

A curious combination of metals has lately very much attracted the attention of the scientific world: I mean the meteoric stones that fall from the atmosphere. They consist principally of native or pure iron, which is never found in that state in the bowels of the earth; and contain also a small quantity of nickel and chrome, a combination likewise new in the mineral kingdom.

These circumstances have led many scientific persons to believe that those substances have fallen from the moon, or some other planet, while others are of opinion either that they are formed in the atmosphere, or are projected into it by some unknown volcano on the surface of our globe.

CAROLINE.

I have heard much of these stones, but I believe many people are of opinion that they are formed on the surface of the earth, and laugh at their pretended celestial origin.

MRS. B.

The fact of their falling is so well ascertained, that I think no person who has at all investigated the subject, can now entertain any doubt of it. Specimens of these stones have been discovered in all parts of the world, and to each of them some tradition or story of its fall has been found connected. And as the analysis of all those specimens affords precisely the same results, there is strong reason to conjecture that they all proceed from the same source. It is to Mr. Howard that philosophers are indebted for having first analysed these stones, and directed their attention to this interesting subject.

CAROLINE.

But pray, Mrs. B., how can solid masses of iron and nickel be formed from the atmosphere, which consists of the two airs, nitrogen and oxygen?

MRS. B.

I really do not see how they could, and think it much more probable that they fall from the moon. — But we must not suffer this digression to take up too much of our time.

The combinations of metals with each other are called alloys; thus brass is an alloy of copper and zinc; bronze, of copper and tin, &c.

EMILY.

And is not pewter also a combination of metal?

MRS. B.

It is. The pewter made in this country is mostly composed of tin, with a very small proportion of zinc and lead.

CAROLINE.

Block-tin is a kind of pewter, I believe?

MRS. B.

Properly speaking, block-tin means tin in blocks, or square massive ingots; but in the sense in which it is used by ignorant workmen, it is iron plated with tin, which renders it more durable, as tin will not so easily rust. Tin alone, however, would be too soft a metal to be worked for common use, and all tin-vessels and utensils are in fact made of plates of iron, thinly coated with tin, which prevents the iron from rusting.

CAROLINE.

Say rather *oxydating*, Mrs. B. — Rust is a word that should be exploded in chemistry.

MRS. B.

Take care, however, not to introduce the word oxydate, instead of rust, in general conversation; for you would probably not be understood, and you might be suspected of affectation.

Metals differ very much in their affinity for each other; some will not unite at all, others readily combine together, and on this property of metals the art of *soldering* depends.

EMILY.

What is soldering?

MRS. B.

It is joining two pieces of metal together, by a more fusible metal interposed between them. Thus tin is a solder for lead; brass, gold, or silver, are solder for iron, &c.

CAROLINE.

And is not *plating* metals something of the same nature?

MRS. B.

In the operation of plating, two metals are united, one being covered with the other, but without the intervention of a third; iron or copper may thus be covered with gold or silver.

EMILY.

Mercury appears to me of a very different nature from the other metals.

MRS. B.

One of its greatest peculiarities is, that it retains a fluid state at the temperature of the atmosphere. All metals are fusible at different degrees of heat, and they have likewise each the property of freezing or becoming solid at a certain fixed temperature. Mercury congeals only at seventy-two degrees below the freezing point.

EMILY.

That is to say, that in order to freeze, it requires a temperature of seventy-two degrees colder than that at which water freezes.

MRS. B.

Exactly so.

CAROLINE.

But is the temperature of the atmosphere ever so low as that?

MRS. B.

Yes, often in Siberia; but happily never in this part of the globe. Here, however, mercury may be congealed by artificial cold; I mean such intense cold as can be produced by some chemical mixtures, or by the rapid evaporation of ether under the air-pump.*

CAROLINE.

And can mercury be made to boil and evaporate?

MRS. B.

Yes, like any other liquid; only it requires a much greater degree of heat. At the temperature of six hundred degrees, it begins to boil and evaporate like water.

Mercury combines with gold, silver, tin, and with several other metals; and, if mixed with any of them in a sufficient proportion, it penetrates the solid metal, softens it, loses its own fluidity, and forms an *amalgam*, which is the name given to the combination of any metal with mercury, forming a substance more or less solid, according as the mercury or the other metal predominates.

EMILY.

In the list of metals there are some whose names I have never before heard mentioned.

MRS. B.

Besides those which Sir H. Davy has obtained, there are several that have been recently disco-

* By a process analogous to that described, page 155. of this volume.

vered, whose properties are yet but little known, as for instance, titanium, which was discovered by the Rev. Mr. Gregor, in the tin-mines of Cornwall; columbium or tantalium, which has lately been discovered by Mr. Hatchett; and osmium, iridium, palladium, and rhodium, all of which Dr. Wollaston and Mr. Tennant found mixed in minute quantities with crude platina, and the distinct existence of which they proved by curious and delicate experiments.

CAROLINE.

Arsenic has been mentioned amongst the metals, I had no notion that it belonged to that class of bodies, for I had never seen it but as a powder, and never thought of it but as a most deadly poison.

MRS. B.

In its pure metallic state, I believe, it is not so poisonous; but it has such a great affinity for oxygen, that it absorbs it from the atmosphere at its natural temperature: you have seen it, therefore, only in its state of oxyd, when, from its combination with oxygen, it has acquired its very poisonous properties.

CAROLINE.

Is it possible that oxygen can impart poisonous qualities? That valuable substance which produces light and fire, and which all bodies in nature are so eager to obtain?

MRS. B.

Most of the metallic oxyds are poisonous, and derive this property from their union with oxygen. The white lead, so much used in paint, owes its pernicious effects to oxygen. In general, oxygen, in a concrete state, appears to be particularly destructive in its effects on flesh or any animal matter; and those oxyds are most caustic that have an acrid burning taste, which proceeds from the metal having but a slight affinity for oxygen, and therefore easily yielding it to the flesh, which it corrodes and destroys.

EMILY.

What is the meaning of the word *caustic*, which you have just used?

MRS. B.

It expresses that property which some bodies possess, of disorganizing and destroying animal matter, by operating a kind of combustion, or at least a chemical decomposition. You must often have heard of caustic used to burn warts, or other animal excrescences; most of these bodies owe their destructive power to the oxygen with which they are combined. The common caustic, called *lunar caustic*, is a compound formed by the union of nitric acid and silver; and it is supposed to owe its caustic qualities to the oxygen contained in the nitric acid.

CAROLINE.

But, pray, are not acids still more caustic than oxyds, as they contain a greater proportion of oxygen?

MRS. B.

Some of the acids are; but the caustic property of a body depends not only upon the quantity of oxygen which it contains, but also upon its slight affinity for that principle, and the consequent facility with which it yields it.

EMILY.

Is not this destructive property of oxygen accounted for?

MRS. B.

It proceeds probably from the strong attraction of oxygen for hydrogen; for if the one rapidly absorb the other from the animal fibre, a disorganisation of the substance must ensue.

EMILY.

Caustics are, then, very properly said to *burn* the flesh, since the combination of oxygen and hydrogen is an actual combustion.

CAROLINE.

Now, I think, this effect would be more properly termed an oxydation, as there is no disengagement of light and heat.

MRS. B.

But there really is a sensation of heat produced by the action of caustics.

EMILY.

If oxygen is so caustic, why does not that which is contained in the atmosphere burn us?

MRS. B.

Because it is in a gaseous state, and has a greater attraction for its electricity than for the hydrogen of our bodies. Besides, should the air be slightly caustic, we are in a great measure sheltered from its effects by the skin; you know how much a wound, however trifling, smarts on being exposed to it.

CAROLINE.

It is a curious idea, however, that we should live in a slow fire. But, if the air was caustic, would it not have an acrid taste?

MRS. B.

It possibly may have such a taste; though in so

slight a degree, that custom has rendered it insensible.

CAROLINE.

And why is not water caustic? When I dip my hand into water, though cold, it ought to burn me from the caustic nature of its oxygen.

MRS. B.

Your hand does not decompose the water; the oxygen in that state is much better supplied with hydrogen than it would be by animal matter, and if its causticity depend on its affinity for that principle, it will be very far from quitting its state of water to act upon your hand. You must not forget that oxyds are caustic in proportion as the oxygen adheres slightly to them.

EMILY.

Since the oxyd of arsenic is poisonous, its acid, I suppose, is fully as much so?

MRS. B.

Yes; it is one of the strongest poisons in nature.

EMILY.

There is a poison called *verdigris*, which forms on brass and copper when not kept very clean; and this, I have heard, is an objection to these metals being made into kitchen utensils. Is this poison likewise occasioned by oxygen?

MRS. B.

It is produced by the intervention of oxygen; for verdigris is a compound salt formed by the union of vinegar and copper; it is of a beautiful green colour, and much used in painting.

EMILY.

But, I believe, verdigris is often formed on copper when no vinegar has been in contact with it.

MRS. B.

Not real verdigris, but compound salts, somewhat resembling it, may be produced by the action of any acid on copper.

The solution of copper in nitric acid, if evaporated, affords a salt which produces an effect on tin that will surprise you, and I have prepared some from the solution we made before, that I might show it to you. I shall first sprinkle some water on this piece of tin-foil, and then some of the salt. — Now observe that I fold it up suddenly, and press it into one lump.

CAROLINE.

What a prodigious vapour issues from it — and sparks of fire I declare!

MRS. B.

I thought it would surprise you. The effect, however, I dare say you could account for, since it is merely the consequence of the oxygen of the salt rapidly entering into a closer combination with the tin.

There is also a beautiful green salt too curious to be omitted; it is produced by the combination of cobalt with muriatic acid, which has the singular property of forming what is called *sympathetic ink.* Characters written with this solution are invisible when cold, but when a gentle heat is applied, they assume a fine bluish green colour.

CAROLINE.

I think one might draw very curious landscapes with the assistance of this ink; I would first make a water-colour drawing of a winter-scene, in which the trees should be leafless, and the grass scarcely green: I would then trace all the verdure with the invisible ink, and whenever I chose to create spring, I should hold it before the fire, and its warmth would cover the landscape with a rich verdure.

MRS. B.

That will be a very amusing experiment, and I advise you by all means to try it.

Before we part, I must introduce to your acquaintance the curious metals which Sir H. Davy has recently discovered. The history of these extraordinary bodies is yet so much in its infancy, that I shall confine myself to a very short account of them; it is more important to point out to you the vast, and apparently inexhaustible, field of research which has been thrown open to our view by Sir H. Davy's memorable discoveries, than to enter into a minute account of particular bodies or experiments.

CAROLINE.

But I have heard that these discoveries, however splendid and extraordinary, are not very likely to prove of any great benefit to the world, as they are rather objects of curiosity than of use.

MRS. B.

Such may be the illiberal conclusions of the ignorant and narrow-minded; but those who can duly estimate the advantages of enlarging the sphere of science, must be convinced that the acquisition of every new fact, however unconnected it may at first appear with practical utility, must ultimately prove beneficial to mankind. But these remarks are scarcely applicable to the present subject; for some of the new metals have already proved eminently useful as chemical agents, and are likely soon to be employed in the arts. For the enumeration of these metals, I must refer you to our list of simple bodies; they are derived from the alkalies, the earths, and three of the acids, all of which had been hitherto considered as undecompoundable or simple bodies.

When Sir H. Davy first turned his attention to the effects of the Voltaic battery, he tried its power on a variety of compound bodies, and gradually brought to light a number of new and interesting facts, which led the way to more important discoveries. It would be highly interesting to trace his steps in this new department of science, but it would lead us too far from our principal object. A general view of his most remarkable discoveries is all that I can aim at, or that you could, at present, understand.

The facility with which compound bodies yielded to the Voltaic electricity, induced him to make trial of its effects on substances hitherto considered as simple, but which he suspected of being compound, and his researches were soon crowned with the most complete success.

The body which he first submitted to the Voltaic battery, and which had never yet been decomposed, was one of the fixed alkalies, called potash. This substance gave out an elastic fluid at the positive wire, which was ascertained to be oxygen, and at the negative wire, small globules of a very high metallic lustre, very similar in appearance to mercury; thus proving that potash, which had hitherto been considered as a simple incombustible body, was in fact a metallic oxyd; and that its incombustibility proceeded from its being already combined with oxygen.

EMILY.

I suppose the wires used in this experiment were of platina, as they were when you decomposed water; for if of iron, the oxygen would have combined with the wire, instead of appearing in the form of gas.

MRS. B.

Certainly: the metal, however, would equally have been disengaged. Sir H. Davy has distinguished this new substance by the name of POTASSIUM, which is derived from that of the alkali, from which it is procured. I have some small pieces of it in this phial, but you have already seen it, as it is the metal which we burnt in contact with sulphur.

EMILY.

What is the liquid in which you keep it?

MRS. B.

It is naptha, a bituminous liquid, with which I shall hereafter make you acquainted. It is almost the only fluid in which potassium can be preserved, as it contains no oxygen, and this metal has so powerful an attraction for oxygen, that it will not only absorb it from the air, but likewise from water, or any body whatever that contains it.

EMILY.

This, then, is one of the bodies that oxydates spontaneously without the application of heat?

MRS. B.

Yes; and it has this remarkable peculiarity that it attracts oxygen much more rapidly from water than from air; so that when thrown into water, however cold, it actually bursts into flame. I shall now throw a small piece, about the size of a pin's head, on this drop of water.

CAROLINE.

It instantaneously exploded, producing a little flash of light! this is, indeed, a most curious substance!

MRS. B.

By its combustion it is reconverted into potash; and as potash is now decidedly a compound body, I shall not enter into any of its properties till we have completed our review of the simple bodies; but we may here make a few observations on its basis, potassium. If this substance is left in contact with air, it rapidly returns to the state of potash, with a disengagement of heat, but without any flash of light.

EMILY.

But is it not very singulr that it should burn better in water than in air?

CAROLINE.

I do not think so: for if the attraction of potassium for oxygen is so strong that it finds no more difficulty in separating it from the hydrogen in water, than in absorbing it from the air, it will no doubt be more amply and rapidly supplied by water than by air.

MRS. B.

That cannot, however, be precisely the reason, for when potassium is introduced under water, without contact of air, the combustion is not so rapid, and indeed, in that case, there is no luminous appearance; but a violent action takes place, much heat is excited, the potash is regenerated, and hydrogen gas is evolved.

Potassium is so eminently combustible, that instead of requiring, like other metals, an elevation of temperature, it will burn rapidly in contact with water, even below the freezing point. This you may witness by throwing a piece on this lump of ice.

CAROLINE.

It again exploded with flame, and has made a deep hole in the ice.

MRS. B.

This hole contains a solution of potash; for the alkali being extremely soluble, disappears in the water at the instant it is produced. Its presence, however, may be easily ascertained, alkalies having the property of changing paper, stained with turmeric, to a red colour; if you dip one end of this slip of paper into the hole in the ice you will see it change colour, and the same, if you wet it with the drop of water in which the first piece of potassium was burnt.

CAROLINE.

It has indeed changed the paper from yellow to red.

MRS. B.

This metal will burn likewise in carbonic acid gas, a gas that had always been supposed incapable of supporting combustion, as we were unacquainted with any substance that had a greater attraction for oxygen than carbon. Potassium, however, readily decomposes this gas, by absorbing its oxygen, as I shall show you. This retort is filled with carbonic acid gas. — I will put a small piece of potassium in it; but for this combustion a slight elevation of temperature is required, for which purpose I shall hold the retort over the lamp.

CAROLINE.

Now it has taken fire, and burns with violence! It has burst the retort.

MRS. B.

Here is the piece of regenerated potash; can you tell me why it is become so black?

EMILY.

No doubt it is blackened by the carbon, which, when its oxygen entered into combination with the potassium, was deposited on its surface.

MRS. B.

You are right. This metal is perfectly fluid at the temperature of one hundred degrees; at fifty degrees it is solid, but soft and malleable; at thirty-two degrees it is hard and brittle, and its fracture exhibits an appearance of confused crystallization. It is scarcely more than half as heavy as water; its specific gravity being about six when water is reckoned at ten; so that this metal is actually lighter than any known fluid, even than ether.

Potassium combines with sulphur and phosphorus, forming sulphurets and phosphurets; it likewise forms alloys with several metals, and amalgamates with mercury.

EMILY.

But can a sufficient quantity of potassium be obtained, by means of the Voltaic battery, to admit of all its properties and relations to other bodies being satisfactorily ascertained?

MRS. B.

Not easily; but I must not neglect to inform you that a method of obtaining this metal in considerable quantities has since been discovered. Two eminent French chemists, Thenard and Gay Lussac, stimulated by the triumph which Sir H. Davy had obtained, attempted to separate potassium from its combination with oxygen, by common chemical means, and without the aid of electricity. They caused red hot potash in a state of fusion to filter through iron turnings in an iron tube, heated to whiteness. Their experiment was crowned with the most complete success; more potassium was obtained by this single operation, that could have been collected in many weeks by the most diligent use of the Voltaic battery.

EMILY.

In this experiment, I suppose, the oxygen quitted its combination with the potassium to unite with the iron turnings?

MRS. B.

Exactly so; and the potassium was thus obtained in its simple state. From that time it has become a most convenient and powerful instrument of deoxygenation in chemical experiments. This important improvement, engrafted on Sir H. Davy's previous discoveries, served but to add to his glory, since the facts which he had established, when possessed of only a few atoms of this curious substance, and the accuracy of his analytical statements, were all confirmed when an opportunity occurred of repeating his experiments upon this substance, which can now be obtained in unlimited quantities.

CAROLINE.

What a satisfaction Sir H. Davy must have felt, when by an effort of genius he succeeded in bringing to light and actually giving existence, to these curious bodies, which without him might perhaps have ever remained concealed from our view !

MRS. B.

The next substance which Sir H. Davy submitted to the influence of the Voltaic battery was *Soda*, the other fixed alkali, which yielded to the same powers of decomposition; from this alkali too, a metallic substance was obtained, very analogous in its properties to that which had been discovered in potash; Sir H. Davy has called it SODIUM. It is rather heavier than potassium, though considerably lighter than water; it is not so easily fusible as potassium.

Encouraged by these extraordinary results, Sir H. Davy next performed a series of beautiful experiments on *Ammonia*, or the volatile alkali, which, from analogy, he was led to suspect might also contain oxygen. This he soon ascertained to be the fact, but he has not yet succeeded in obtaining the basis of ammonia in a separate state; it is from analogy, and from the power which the volatile alkali has, in its gaseous form, to oxydate iron, and also from the amalgams which can be obtained from ammonia by various processes, that the proofs of that alkali being also a metallic oxyd are deduced.

Thus, then, the three alkalies, two of which had always been considered as simple bodies, have now lost all claim to that title, and I have accordingly classed the alkalies amongst the compounds, whose properties we shall treat of in a future conversation.

EMILY.

What are the other newly discovered metals which you have alluded to in your list of simple bodies?

MRS. B.

They are the metals of the earths which became next the object of Sir H. Davy's researches; these bodies had never yet been decomposed, though they were strongly suspected not only of being compounds, but of being metallic oxyds. From the circumstance of their incombustibility it was conjectured, with some plausibility, that they might possibly be bodies that had been already burnt.

CAROLINE.

And metals, when oxydated, become, to all appearance, a kind of earthy substance.

MRS. B.

They have, besides, several features of resemblance with metallic oxyds; Sir H. Davy had therefore great reason to be sanguine in his expectations of decomposing them, and he was not disappointed. He could not, however, succeed in obtaining the basis of the earths in a pure separate state; but metallic alloys were formed with other metals, which sufficiently proved the existence of the metallic basis of the earths.

The last class of new metallic bodies which Sir H. Davy discovered was obtained from the three undecompounded acids, the boracic, the fluoric, and the muriatic acids; but as you are entirely unacquainted with these bodies, I shall reserve the account of their decomposition till we come to treat of their properties as acids.

Thus in the course of two years, by the unparalleled exertions of a single individual, chemical science has assumed a new aspect. Bodies have been brought to light which the human eye never before beheld, and which might have remained eternally concealed under their impenetrable disguise.

It is impossible at the present period to appreciate to their full extent the consequences which science or the arts may derive from these discoveries; we may, however, anticipate the most important results.

In chemical analysis we are now in possession of more energetic agents of decomposition than were ever before known.

In geology new views are opened, which will probably operate a revolution in that obscure and difficult science. It is already proved that all the earths, and, in fact, the solid surface of this globe, are metallic bodies mineralized by oxygen, and as our planet has been calculated to be considerably more dense upon the whole than on the surface, it is reasonable to suppose that the interior part is composed of a metallic mass, the surface of which only has been mineralized by the atmosphere.

The eruptions of volcános, those stupendous problems of nature, admit now of an easy explanation. For if the bowels of the earth are the grand recess of these newly discovered inflammable bodies, whenever water penetrates into them, combustions and explosions must take place; and it is remarkable that the lava which is thrown out, is the very kind of substance which might be expected to result from these combustions.

I must now take my leave of you; we have had a very long conversation to-day, and I hope you will be able to recollect what you have learnt. At our next interview we shall enter on a new subject.

END OF THE FIRST VOLUME.

Printed by A. Strahan,
Printers-Street, London.

Printed in the United States
By Bookmasters